Nurturing and Sustaining Effective Programs in Science Education for Grades K–8

BUILDING A VILLAGE IN CALIFORNIA

Summary of a Convocation

Steve Olson, *Rapporteur*
Jay B. Labov, *Editor*

NATIONAL ACADEMY OF SCIENCES *AND*
NATIONAL ACADEMY OF ENGINEERING
OF THE NATIONAL ACADEMIES

THE NATIONAL ACADEMIES PRESS
Washington, D.C.
www.nap.edu

THE NATIONAL ACADEMIES PRESS   500 Fifth Street, N.W.   Washington, DC 20001

NOTICE: The project that is the subject of this report was approved by the Governing Board of the National Research Council, whose members are drawn from the councils of the National Academy of Sciences, the National Academy of Engineering, and the Institute of Medicine.

This study was supported by the Arnold and Mabel Beckman Fund of the National Academy of Sciences and the National Academy of Engineering and a grant from the S.D. Bechtel, Jr. Foundation. Any opinions, findings, and conclusions or recommendations expressed in this publication are those of the persons identified in the report and do not necessarily reflect the views of the National Academies.

International Standard Book Number-13:   978-0-309-14366-0
International Standard Book Number-10:   0-309-14366-7

Additional copies of this report are available from the National Academies Press, 500 Fifth Street, N.W., Lockbox 285, Washington, DC 20055; (800) 624-6242 or (202)334-3313 (in the Washington metropolitan area); Internet, http://www.nap.edu.

Cover credit: Stock photos from Getty Images®.

Suggested citation: National Academy of Sciences and National Academy of Engineering. (2009). *Nurturing and Sustaining Effective Programs in Science Education for Grades K-8: Building a Village in California: Summary of a Convocation.* Steve Olson, Rapporteur. Jay B. Labov, Editor. Washington, DC: The National Academies Press.

THE NATIONAL ACADEMIES
Advisers to the Nation on Science, Engineering, and Medicine

The **National Academy of Sciences** is a private, nonprofit, self-perpetuating society of distinguished scholars engaged in scientific and engineering research, dedicated to the furtherance of science and technology and to their use for the general welfare. Upon the authority of the charter granted to it by the Congress in 1863, the Academy has a mandate that requires it to advise the federal government on scientific and technical matters. Dr. Ralph J. Cicerone is president of the National Academy of Sciences.

The **National Academy of Engineering** was established in 1964, under the charter of the National Academy of Sciences, as a parallel organization of outstanding engineers. It is autonomous in its administration and in the selection of its members, sharing with the National Academy of Sciences the responsibility for advising the federal government. The National Academy of Engineering also sponsors engineering programs aimed at meeting national needs, encourages education and research, and recognizes the superior achievements of engineers. Dr. Charles M. Vest is president of the National Academy of Engineering.

The **Institute of Medicine** was established in 1970 by the National Academy of Sciences to secure the services of eminent members of appropriate professions in the examination of policy matters pertaining to the health of the public. The Institute acts under the responsibility given to the National Academy of Sciences by its congressional charter to be an adviser to the federal government and, upon its own initiative, to identify issues of medical care, research, and education. Dr. Harvey V. Fineberg is president of the Institute of Medicine.

The **National Research Council** was organized by the National Academy of Sciences in 1916 to associate the broad community of science and technology with the Academy's purposes of furthering knowledge and advising the federal government. Functioning in accordance with general policies determined by the Academy, the Council has become the principal operating agency of both the National Academy of Sciences and the National Academy of Engineering in providing services to the government, the public, and the scientific and engineering communities. The Council is administered jointly by both Academies and the Institute of Medicine. Dr. Ralph J. Cicerone and Dr. Charles M. Vest are chair and vice chair, respectively, of the National Research Council.

www.national-academies.org

CONVOCATION ORGANIZERS

Bruce Alberts,[*] Professor, University of California, San Francisco

Maureen Allen, Consultant/Instructor, Beckman@Science, Los Alamitos

Raymond Bartlett, Senior STEM Consultant, Teaching Institute for Essential Science, Columbia, Maryland

Eilene Cross, Consultant, California Council on Science and Technology, Pleasanton

Angela Phillips Diaz, Special Assistant to the Chancellor, University of California, Riverside

Jacqueline Dorrance, Executive Director, Arnold and Mabel Beckman Foundation, Irvine

Susan Elrod, Director, Center for Excellence in Science and Mathematics Education, California Polytechnic State University

Susan Hackwood, Executive Director, California Council on Science and Technology, Riverside

Susan Harvey, Senior Program Officer, S.D. Bechtel, Jr. Foundation, San Francisco

Harry Helling, President and Chief Executive Officer, Crystal Cove Alliance, Newport Coast

Michael Masterson, Systems and Research Manager, California STEM Innovation Network, California Polytechnic State University

Sue Neuen, Director of Professional Development, California Science Center, Los Angeles

Soo Venkatesan, Program Officer, S.D. Bechtel, Jr. Foundation, San Francisco

Staff

Jay Labov (*Convocation Director*), Senior Advisor for Education and Communication, National Academy of Sciences and National Research Council

Dorothy Majewski, Senior Project Assistant, Center for Education, National Research Council

---

[*]Member, National Academy of Sciences.

Contents

Appendixes

Preface

Science education in California is in a state of crisis. On the 2005 National Assessment of Educational Progress (NAEP), California eighth graders ranked second lowest among the students of all states in science. On the 2007 fifth grade California Standards Test in Science, only 37 percent of students scored at a proficient level, even though proficiency levels are set lower in California than for the NAEP test. In a recent survey of San Francisco Bay Area elementary school teachers, four of five teachers in kindergarten through fifth grade reported spending less than an hour on science each week; one in six teachers devoted no time at all to science. And all these problems are likely to intensify: an ongoing fiscal crisis in the state threatens further cutbacks, teacher and administrator layoffs, and less money for professional development.

It is especially painful to watch the deterioration of science education in California given the increasingly pervasive roles that science and technology play in the state. For decades, California has led the nation and the world through innovations in aerospace, energy production, digital technologies, biotechnology, agriculture, environmental technologies, and multimedia entertainment. Of the nation's top 10 research universities (as measured by federal research funding), 4 are in California. Almost a third of the members of the National Academy of Sciences and more than a fourth of the members of the National Academy of Engineering live and work in California. The human and institutional infrastructure needed to rebuild science education in California is available. But the problems plaguing K-12 education in California have kept many teachers, schools, and school districts from taking advantage of the state's rich resources in science and technology.

On April 29-30, 2009, the National Academy of Sciences and the National Academy of Engineering, in association with the California Council on Science and Technology (CCST), the Arnold and Mabel Beckman Foundation, and the S.D. Bechtel, Jr. Foundation, hosted a meeting at the Arnold and Mabel Beckman Center in Irvine, California, to confront the crisis in California science education, particularly at the kindergarten through eighth grade level. The convocation brought together key stakeholders in the science education system to enable and facilitate an exploration of ways to more effectively and efficiently support, sustain, and communicate across the state concerning promising research and practices in K-8 science education. Through discussions organized and moderated by facilitators, convocation participants examined data collected from several large science education initiatives, one in California and two in other states (Washington and New Jersey), that have been deemed successful on the basis of various kinds of evidence. Convocation participants were tasked with considering what additional kinds of research might be valuable as part of efforts like these.

Importantly, the organizers of the convocation made clear to all attendees that the convocation was not a one-time event. Rather, it was specifically designed to launch a series of subsequent activities aimed at developing a workable model for sustaining effective education programs in California. These are to be overseen by CCST and other organizations in the state.

The meeting was remarkable for the diversity of institutions represented, including the state and federal governments, business and industry, K-12 teachers and administrators, higher education, philanthropic organizations, and education researchers. Many participants observed that the meeting was their first opportunity to talk at length and in depth with such a wide-ranging group of stakeholders in the California education system.

The current state of affairs is unacceptable, many participants said. Ways must be found to harness the existing base of experience, knowledge, and research-based evidence about effective education programs for *all* students to return science education in California to world-class status. All of the sectors with an interest in science education must agree on what needs to be done and then work together to achieve those objectives. While each group has an essential and unique role to play, improvement of science and technology education, especially at the K-8 level, will be possible only when all stakeholders share a common vision and goals.

An idea that was extensively discussed at the workshop and described in Chapter 6 of this summary of the meeting is the formation of a broad-based coalition designed to reform science education in California. Efforts to build such a coalition have continued since the meeting, and this report offers ideas about how those efforts can continue to progress and how new initiatives can foster and support more broadly based efforts to improve science, technology, engineering, and mathematics education. But time is short. Once a base of expertise among K-8 teachers and administrators is lost, rebuilding that base can be very difficult.

A few days before the convocation, President Barack Obama spoke at the headquarters of the National Academy of Sciences in Washington, DC. Observing that "the progress and prosperity of future generations will depend on what we do now to educate the next generation," he announced "a renewed commitment to education in mathematics and science." The challenges are daunting, he said, but America has risen to such challenges many times before. "Even in the hardest times and against the toughest odds, we have never given in to pessimism, we have never surrendered our fates to chance, we have worked hard, we have sought out new frontiers."[1]

> *The progress and prosperity of future generations will depend on what we do now to educate the next generation.*
> —President Barack Obama

Throughout its history, California has led the nation into the future. It has pioneered many of the innovations that have made the United States a world leader in industry and culture. It has served as a beacon of American energy and optimism. In the midst of the current crisis, with the support and creativity of the many people and organizations that both influence and depend on effective science education for students in grades K-8 and beyond, California has an opportunity to again demonstrate the clear-sighted vision that has resulted in so many past triumphs. We hope that the convocation marked a turning point and that this publication will help a much broader group of stakeholders better understand and act in concert to ensure an effective science education for all students in California and throughout the nation.

Bruce Alberts
Jay Labov

---

[1]Video and audio streaming and a transcript of President Obama's speech are available at http://www.nationalacademies.org/morenews/20090428.html.

Structure of the Report

This summary provides a narrative, rather than a chronological, overview of the presentations and the ensuing rich discussions that permeated the convocation. It brings together into individual chapter's themes that were raised and recurred throughout the event.

- Chapter 1, The Challenges Facing California, focuses on introductory remarks that set the theme and the tone for the entire convocation about the multitude of challenges that are currently besetting science education in California, especially attempts to educate children in Grades K-8.
- Chapter 2, The National Context, describes presentations and discussion that helped to frame the current environment in California with concerns and science education initiatives nationally.
- Members of the organizing group were concerned that some participants might never have experienced hands-on, inquiry-based approaches to science education. Thus, part of the morning of Day 1 was devoted to demonstrations of such approaches with elementary school children from Orange County. Chapter 3, Science Education in Action, describes these activities.
- Chapter 4, Exemplary Programs, focuses on presentations from Day 1 about three programs for K-8 science education considered by the organizing group to be exemplary based on evidence that has been systematically collected from their inception. These pre-

sentations served as the basis for further discussion during the afternoon breakout sessions on Day 1.

- Day 2 was devoted to synthesizing information from the presentations, plenary discussions, and breakout sessions from Day 1. Chapter 5, Fostering Sustainable Programs, describes these very rich conversations and how they were used as preludes to the final breakout sessions of the morning. The breakout sessions enabled people from each of the sectors represented to meet with each other, discuss what they had heard, and begin to formulate plans for action that their sector could most effectively promote and execute.
- Chapter 6, Rising to the Challenge, is devoted to reporting the ideas and plans for action that each of the sectors had developed. This chapter is now serving as the basis for subsequent activities that have begun to unfold across California.

Throughout the report, footnotes provide links to websites of all the organizations and programs mentioned in the narrative. The References section contains complete citations to the data that were discussed during the presentations.

Appendixes A and B provide readers with the convocation agenda and a list of participants, respectively. Appendix C contains biographical sketches of the convocation organizers and presenters. Appendix D presents summaries of 10 major reports on various aspects of science, technology, engineering, and mathematics education published by the National Academies during the past decade. Each report was selected for inclusion for its direct implications for the topics discussed at the convocation.

Acknowledgments

This publication summarizes the presentations made at the Convocation on Sustaining Effective Science Education Programs for Grades K-8 in California. The summary is limited to material presented and statements made by individual participants at the convocation.

This convocation summary has been reviewed in draft form by individuals chosen for their diverse perspectives and technical expertise. The purpose of this independent review is to provide candid and critical comments that will assist the institution in making its published report as sound as possible and to ensure that the summary meets institutional standards for objectivity, evidence, and responsiveness to the charge. The review comments and draft manuscript remain confidential to protect the integrity of the process. We thank the following individuals for their careful and thoughtful review of this report: Lynn E. Baroff, Office of the Executive Director, California Space Education and Workforce Institute, Pasadena; Richard Cardullo, Department of Biology, University of California, Riverside; Jim Gentile, President, Research Corporation for Science Advancement, Tucson; and Kathryn Scantlebury, Department of Chemistry and Biochemistry, University of Delaware.

Although the reviewers listed above provided many constructive comments and suggestions, they were not asked to endorse the content of the report, nor did they see the final draft of the report before its release. The review of this report was overseen by Carlo Parravano, executive director, Merck Institute for Science Education. He was responsible for making certain that an independent examination of this summary was carried out in accordance with institutional procedures and that all review comments were carefully considered. Responsibility for the final content of this summary rests entirely with the author, the editor, and the institution.

We sincerely thank the Arnold and Mabel Beckman Foundation and the S.D. Bechtel, Jr. Foundation for their support of this convocation and the production and dissemination of this report.

1

The Challenges Facing California

Key Points

- Many indicators point to severe weaknesses in California's science education systems at the kindergarten through eighth grade (K-8) levels.
- K-8 students in California spend too little time studying science, many of their teachers are not well prepared in the subject, and the support system for science instruction has deteriorated.
- A proliferation of overly detailed standards and poorly conceived assessments has trivialized science education.
- Yet there exists a solid base on which to strengthen K-8 science education in California and across the nation, including a nascent movement toward common national standards, new research findings on effective educational practices; the involvement of scientific, business, and philanthropic organizations in many schools; and the growing realization that science education must improve to support future prosperity.

Our children's future will be filled with incredible, advanced technologies—the likes of which we can only dream of today. . . . Science literacy, therefore, will no longer be an advantage, but an absolute necessity for success.
                                                                —Arnold Beckman

The storm that threatens the economic prospects of California—and the rest of the United States—is clearly visible, said Jacqueline Dorrance, executive director of the Arnold and Mabel Beckman Foundation, in her welcoming remarks at the convocation "It Takes a Village: Sustaining Effective Education Programs in Science for Grades K-8."

- California has ranked near the bottom of all states in the percentage of fourth graders at or above proficiency in science (see Figure 1-1).
- According to a national poll conducted by the Bayer Corporation (1995), 68 percent of parents and 64 percent of elementary school teachers do not consider themselves to be scientifically literate.
- In international tests conducted as part of the 2006 Program for International Student Assessment (PISA),[1] U.S. 15-year-olds ranked 25th out of 30 countries in mathematics and 21st in science (Organisation for Economic Co-operation and Development, 2007).
- The number of people who speak and are learning English in China is greater than the population of the United States (Yang, 2006).
- The top quarter of students in India outnumbers the total number of students in the United States.[2]
- Between 2005 and 2006 the United States dropped from first to sixth place in the World Economic Forum's index of global economic competitiveness (World Economic Forum, 2006).
- An estimated 14 million U.S. workers (11 percent of the total workforce in 2001 at the time when the estimate was made) currently occupy jobs that have the potential to be outsourced to other countries (Bardhan and Kroll, 2003).

SCIENCE EDUCATION IN CALIFORNIA

As the world continues to change at an ever-faster pace, policy leaders in the United States must ask themselves how they can help prepare the nation's children to succeed in an increasingly competitive and technologically advanced workplace, Dorrance said. "Are we to give up our competitive edge? Are we to rely on other countries to fill our scientific workforce?"

> *Are we to give up our competitive edge? Are we to rely on other countries to fill our scientific workforce?*
>
> —Jacqueline Dorrance

---

[1]For additional information, see http://www.pisa.oecd.org/pages/0,2987,en_32252351_32235731_1_1_1_1,00.html.

[2]For additional information, see http://www.youtube.com/watch?v=K04o2ic4g-A.

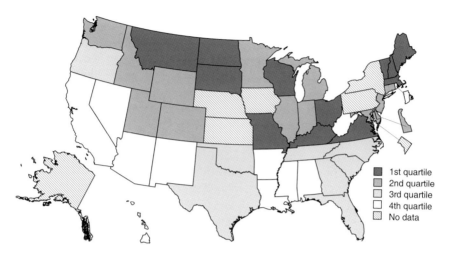

FIGURE 1-1 In 2005, California fourth graders ranked in the lowest quartile of U.S. states in science proficiency.
SOURCE: National Science Board (2008, pp. 8-14).

Over the course of the convocation, which took place on April 29-30, 2009, at the Arnold and Mabel Beckman Center in Irvine, California, other speakers pointed to the many ways in which science education in elementary schools, middle schools, and junior high schools is failing to prepare California students for the future. Rena Dorph summarized a study of science education in elementary schools in the San Francisco Bay Area conducted by the Center for Research, Evaluation, and Assessment at the University of California at Berkeley's Lawrence Hall of Science and WestEd (Dorph et al., 2007). The study collected data from a wide variety of sources, including districts, teachers, and science-rich education institutions, such as science centers, museums, and laboratories. Although the Bay Area is a center of innovation in science and technology in the United States, school districts there face challenges similar to those faced across the state. The study shows that 1 in 7 Bay Area teachers has been teaching less than two years, and in 16 Bay Area districts (representing 20 percent of Bay Area students) 20 to 35 percent of the teachers fall into this category, including many of the districts that serve the largest populations. In spring 2007, just 46 percent of Bay Area fifth graders scored at a proficient level or above on the California Standards Test in science, which is slightly better than the 37 percent of all California fifth graders who scored at this level, but still alarmingly low.

The amount of time spent on science in Bay Area elementary schools is very limited, Dorph pointed out. According to teacher surveys, 80 percent of K-5 multiple-subject teachers spend 60 minutes or less on science

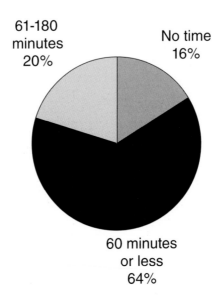

61-180
minutes
20%

No time
16%

60 minutes
or less
64%

FIGURE 1-2 A survey of Bay Area teachers in self-contained K-5 classrooms showed that only one in five spends more than an hour on science per week. SOURCE: Dorph et al. (2007). Copyright © 2007 The Regents of the University of California. Used with permission.

per week (see Figure 1-2). One in six multiple-subject elementary school teachers reports spending no time at all on science. Teacher and district surveys indicate that less time has been spent on science since passage of the No Child Left Behind Act of 2001, which initially mandated regular testing in mathematics and reading but not science. In particular, many districts with schools that have been targeted for improvement because of their poor test results in mathematics and reading report spending little or no time on science.

The survey also found that teachers feel less prepared to teach science than they do to teach other subjects and they have few opportunities to improve their preparation. According to the survey of teachers, 41 percent feel unprepared to teach science compared with 4 percent in both mathematics and language arts (see Figure 1-3). A study by the chancellor's office of the California State University system found similar results, said Eilene Cross of the California Council on Science and Technology (CCST). According to that study, only about 40 percent of elementary teachers feel that they have been well prepared to teach science (California State University Center for Teacher Quality, 2008).

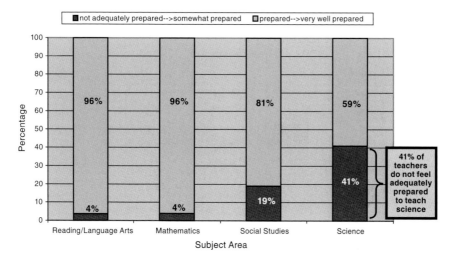

FIGURE 1-3 Many multiple-subject K-5 teachers in the Bay Area feel less adequately prepared to teach science than they do for other subjects.
SOURCE: Dorph et al. (2007). Copyright © 2007 The Regents of the University of California. Used with permission.

Cross is currently engaged in a qualitative study of the gap between standards-based, high-quality elementary science education and the preparation prospective teachers receive in college to teach science. Information is being gathered from the California Commission on Teacher Credentialing,[3] from teacher preparation programs, and from elementary schools. A high-level advisory group of CCST members and representatives from the private sector is providing guidance on research questions, data sources, and presentation of results. In addition, focus groups of educators and others are providing insights into the data and the preliminary findings. The results of the study will inform statewide initiatives in elementary science education. The goal is to "describe what's going on in the classroom and make recommendations for the future," said Cross.

Furthermore, according to the Dorph et al. study, most districts offer minimal professional development in science. Over the past three years, 32 percent of multiple-subject elementary school teachers in self-contained classrooms report receiving fewer than six hours of professional development in science, and 38 percent report receiving none (see Figure 1-4).

---

[3]For additional information about the California Commission on Teacher Credentialing, see http://www.ctc.ca.gov/.

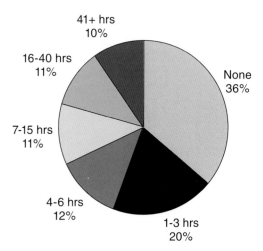

FIGURE 1-4 More than two-thirds of K-5 multiple-subject teachers in self-contained classrooms reported receiving either no or less than six hours of professional development in science over the past three years.
SOURCE: Dorph et al. (2007). Copyright © 2007 The Regents of the University of California. Used with permission.

About half of districts report that they do not have capacity in district offices to support science at the elementary level.

Anne Marie Bergen, a middle school teacher from the Oakdale Joint Unified School District who also serves as chair of the California Teacher Advisory Council for the California Council on Science and Technology,[4] emphasized the lack of preparation many elementary school teachers have received in science. "Many teachers are not comfortable teaching science," she said. "Even if they are excellent teachers, they don't feel competent. They feel . . . insecure with content and intimidated by using materials."

> *Many teachers are not comfortable teaching science. Even if they are excellent teachers, they don't feel competent.*
>
> —Anne Marie Bergen

---

[4]For additional information about the California Teacher Advisory Council, see http://ccst.us/ccstinfo/caltac.php. For additional information about the California Council on Science and Technology, see http://ccst.us.

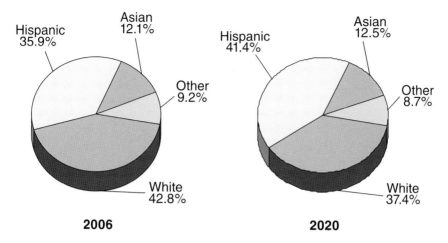

2006   **2020**

FIGURE 1-5 The population of California is becoming ethnically more diverse.
SOURCE: California Budget Project (2008).

Teacher preparation at the university and college level is a major contributor to this problem, she said. Most prospective elementary school teachers do not receive the preparation they need in science, either in college or once they begin teaching. Some teachers seek out professional development to improve their knowledge of science and science instruction, but that is the exception and not the rule.

Elementary school teachers also need supportive schools and communities to be able to teach science well. "The school board needs to understand what good science looks like, your principal really has to understand it," Bergen said. "You also have to have parents who understand it, and to understand it they need to see what it looks like. That's a huge issue."

The student population in California has become increasingly diverse, as is the case in many parts of the United States, and it will become even more so in the future (see Figure 1-5). For example, English language learners constitute a large fraction of the students in many of California's districts. Kathy DiRanna, the statewide director of WestEd's K-12 Alliance,[5] observed that one-third of the nation's English language learners are in California, and they constitute about a quarter of the K-12 students in the state. About 54 percent of students in Orange County's Santa Ana

---

[5]For additional information about the K-12 Science Alliance, see http://www.wested.org/cs/we/view/pj/79. For additional information about WestEd, see http://www.wested.org/cs/we/print/docs/we/home.htm.

Unified School District are English language learners, and they constitute 38 percent of students in the Los Angeles Unified School District
and 28 percent of students in the San Francisco and the San Diego Unified School Districts. Many English language learners are in schools and
districts that lack resources, and they are often in districts with many
beginning teachers. Furthermore, it is a very diverse population, with
great differences in the language they speak, their parents' backgrounds,
whether they are literate in their home language, and the number of years
they have lived in the United States.

Dennis Bartels, the executive director of the Exploratorium in San
Francisco,[6] described a transformation that he observed in California
science education in just the past few years. In 2001, when Bartels left an
earlier position at the Exploratorium to become president of the education
research and development center TERC in Massachusetts, the state had "a
robust system of support for science education and for all of education,"
he said. When he returned to become executive director of the Exploratorium in 2006, "I could not believe how much had changed in just five
years. What the hell happened?"

In addition to the emphasis on mathematics and reading created by
the No Child Left Behind legislation, a series of state budget crises severely
limited the resources devoted to education. Another major change, according to Bartels, was the loss of an extensive system of supports for science teachers. In 1989 the National Science Foundation supported more
than 60 teacher enhancement projects in California, whereas today its
involvement in teacher preparation in California is much reduced. Also,
as recently as 2001, the California Subject Matter Project,[7] a professional
development organization for California educators, was a model for other
states, particularly in the area of science. "The program disappeared over-
night except for some minimal funding for science," said Bartels. In 2007
the state had about one-third of the support dollars for science that it had
just seven years earlier.

The consequences of these many weaknesses in California's science education system are obvious, said Susan Hackwood, the executive
director of CCST. Only 4 percent of ninth graders in California ultimately
graduate from college with a degree in science or engineering (California
Council on Science and Technology, 2002). As many as 60 percent of college freshmen intending to earn such a degree do not do so. "California is
very good at inventing technologies and bringing them to practice," said
Hackwood. But "we're lousy at producing the workforce that is going to
populate the jobs that we create. We rely entirely on bringing in people

---

[6]For additional information, see http://exploratorium.com.
[7]For additional information, see http://csmp.ucop.edu.

from other states and other countries. . . . If we don't have that flow of people coming into California, we're dead."

GOALS FOR SCIENCE EDUCATION

In his opening remarks at the convocation, Bruce Alberts of the University of California, San Francisco, former president of the National Academy of Sciences, and now the current editor-in-chief of *Science*, said that he has three goals for science education:

1. Enable all children to acquire the problem-solving, thinking, and communication skills of scientists so they can be productive and competitive in the new world economy.
2. Foster a scientific temperament for the nation, with scientifically trained people in many professions, to help ensure the rationality and tolerance essential for a democratic society.
3. Help the United States generate new scientific knowledge and technology by casting the widest possible net for talent.

The National Science Education Goals that Alberts promoted as president of the National Academy of Sciences in the 1990s were designed in part to achieve these objectives (National Research Council, 1996). But what followed was often a "disaster," according to Alberts, as individual states produced their own standards for science education and often paid little attention to the national standards. The science standards adopted in California in 1998, for example, were much more detailed than the national standards. The proliferation of different state standards means that textbook publishers had to try to match the needs of multiple states. As a result, they crammed large amounts of material into textbooks in an attempt to meet different state standards. In addition, the diversity of standards greatly complicated the effort to produce high-quality assessments that can help guide instruction.

Alberts cited as an example a seventh grade textbook that devotes the following two sentences to describe the endoplasmic reticulum. "Running through the cell is a network of flat channels called the **endoplasmic reticulum.** This organelle manufactures, stores, and transports materials." At the end of the chapter, a self-test asks that students "write a sentence that uses the term 'endoplasmic reticulum' correctly."

"It's an absolute tragedy," said Alberts. "We're telling kids that education is a joke," a problem exacerbated by the much more engaging electronic media available to students. It is much easier to test students for their familiarity with scientific words than for scientific understanding and abilities. "Bad tests are forcing a trivialization of science and will drive most students, including many potential scientists, away from science," Alberts said. "Any rational person would be totally turned off from school if this is what school is about. It doesn't mean anything."

> *It's an absolute tragedy. We're telling kids that education is a joke.*
> —Bruce Alberts

The National Research Council's report *Taking Science to School*, a comprehensive analysis of how students in grades K-8 can best learn science, concluded that effective science education needs to combine four strands of learning (National Research Council, 2007a). They are as follows:

1. Know, use, and interpret scientific explanations of the natural world.
2. Generate and evaluate scientific evidence and explanations.
3. Understand the nature and development of scientific knowledge.
4. Participate productively in scientific practices and discourse.

Each of these four strands of inquiry was judged by the committee that wrote the report to be of equal importance, Alberts observed. Furthermore, strands 2, 3, and 4 can be taught only through the use of inquiry.

These strands correspond closely with the workforce skills needed by business and industry. According to the book *Thinking for a Living* by Ray Marshall and Marc Tucker (1992), the skills that employers need from high school graduates are a high capacity for abstract conceptual thinking, the ability to apply that capacity for abstract thought to complex, real-world problems, and the capacity to function effectively in an environment in which communication skills are vital in work groups. Inquiry-based science instruction "perfectly meets this challenge," Alberts said. He quoted Robert Galvin, the former chief executive officer of Motorola: "While most descriptions of necessary skills for children do not list 'learning to learn,' this should be the capstone skill upon which all others depend." Memorized facts, which are the basis for most testing done in schools today, are of little use in an age in which information is doubling every two or three years. Computers and the Internet can provide information when it is needed. The workforce needs to know how to use information to develop solutions to problems.

Yet most science education is a far cry from the kind called for in *Taking Science to School*. Here is Alberts's interpretation of the science education that takes place in most U.S. classrooms:

1. Know, use, and interpret scientific explanations of the natural world.
2. Generate and evaluate scientific evidence and explanations.
3. Understand the nature and development of scientific knowledge.
4. Participate productively in scientific practices and discourse.

"We are not doing science education according to the people who actually understand what science education should be," said Alberts. "And it's no wonder that parents don't think much of science education, because what we're doing has defined science education for them, and they don't think it's important. And they're probably right."

Other nations are ahead of the United States in fashioning an effective science education system. The PISA assessment of students' scientific knowledge and skills, for example, is rooted in the concept of scientific literacy, which it defines as the extent to which an individual:

- Possesses scientific knowledge and uses that knowledge to identify questions, acquire new knowledge, explain scientific phenomena, and draw evidence-based conclusions about science-related issues.
- Understands the characteristic features of science as a form of human knowledge and inquiry.
- Shows awareness of how science and technology shape the material, intellectual, and cultural environments.
- Engages in science-related issues and with the ideas of science, as a reflective citizen.

The United States has an opportunity to move toward this vision of science education, said Alberts, "because there's a widespread recognition that the current chaotic system does not work." For example, the National Governors Association and the Council of Chief State School Officers[8] had a meeting in Chicago in April 2009, at which they decided to push for common standards in mathematics and English language arts, with science to follow. The standards would be aligned with college- and career-ready expectations and made available for states to adopt voluntarily.

"Many other promising signs of change are evident," Alberts said. Foundations are putting money into education. A major redesign of Advanced Placement courses in biology, chemistry, and physics is under way. President Barack Obama and his science adviser, John Holdren, have signaled their intention to devote considerable attention to science education. The President's Council of Advisors on Science and Technology, the members of which were announced a few days before the convocation, is establishing a subcommittee on science education.

---

[8]For additional information, see http://ccsso.org.

Alberts also has instituted a major reform at *Science* to raise the visibility of science education at all grade levels. An Education Forum begun by the previous editor is being featured in the journal through special issues, research and news articles, and an editor devoted to education research. "We need to collect this [information] in one place and make it much more visible," he said.

THE PROMISE MOVING FORWARD

Other speakers at the convocation described many ways in which education in California can be strengthened. For example, Dorph described a recent study with which she was involved of science-rich education institutions in the Bay Area, many of which serve schools directly. The study found that 35 percent of Bay Area K-5 public school students are reached through field trips or onsite classes, and about 20 percent are reached through outreach to their schools. These are estimates, said Dorph, "but they help us get an idea of who's being reached and who not." Science-rich education institutions also provide informal and out-of-school learning opportunities for some students who have little or no science in schools. "This is where science comes to life for them, and it's critical to continue to provide these opportunities while we're working on what's going on in schools."

The earlier study of Bay Area elementary schools also found that the majority of elementary schools receive support for science education from an external partner, including informal learning institutions, colleges and universities, local businesses, and community-based organizations. Teachers rate the quality of the professional development they receive from these external sources higher that those in the public school system. "Teachers also indicated a desire for additional professional development from these organizations," said Dorph, and four in five of their professional development programs have the potential to scale up. Funding for scale-up could support additional staff, teacher stipends, release days, and substitute teachers.

Many California elementary and middle school teachers also can take advantage of materials that could be used to improve science education. For example, about half of the state's elementary students are in classrooms that have access to inquiry-based materials, such as those from the Full Option Science System (FOSS) developed by the Lawrence Hall of Science.[9] "That doesn't mean that the teacher opened the box," said

---

[9]Additional information about FOSS kits is available at http://www.fossweb.com/. Information about the Lawrence Hall of Science is available at http://www.lawrencehallofscience.org.

DiRanna. But across the rest of the United States, only about 20 percent of districts use reform-based materials. California has a tremendous opportunity to use such materials in a much larger fraction of its classrooms.

Finally, the introduction of new fifth grade assessments in science presents opportunities to build science education support systems.[10] Many districts are planning to increase the amount of time spent on science. Specific strategies mentioned by convocation participants are to increase the classroom time spent on science, select new materials, integrate science with mathematics or language arts, provide more opportunities for professional development in science, seek new funding sources to support science education, and leverage the expertise of passionate science teachers.

Susan Pritchard, a middle school science teacher with the La Habra City School District and the current president of the California Science Teachers Association,[11] observed that bringing together the people represented at the convocation was an important step. "We have in this room a lot of people with a lot of moxie." When multiple stakeholders speak with a unified voice, positive change can result. She cited as an example a recent legislative initiative to limit the amount of hands-on activity in science classes to less than 25 percent of instructional time. With support from many colleagues, the California Science Teachers Association (CSTA) was able to change the requirement to *no less* than 25 percent. "That's huge, [but] it takes everybody to work on it." Currently the CSTA is calling on the state legislature to revise the state science standards adopted in 1998. Bills that would have required review and revision of the standards were passed by the California legislature in 2005 and 2006 but were vetoed by the governor. Getting such a bill enacted will require that everyone "get on the bandwagon," said Pritchard.

Together these factors provide considerable grounds for optimism that science education in California could be on the verge of turning a corner. The challenge is to use the problems to motivate reform. "We are here today to turn ideas into action," said Dorrance, in summarizing the goals of the convocation.

The United States of America is still the best country in the world. We are a nation of leaders, and we have always had the competitive edge. But times are changing, and we need to be prepared to keep that edge. We need to take our science education system to the next level, to the level that will help us to be responsive to our future needs. I believe that we,

---

[10]For additional information on the California state testing and reporting program, see http://www.ed-data.k12.ca.us/articles/Article.asp?title=Understanding%20the%20STAR.

[11]For additional information, see http://www.cascience.org/csta/csta.asp.

this village, has an extraordinary opportunity to change students' lives. An opportunity to improve our country. Will it be easy? Of course not. Will we have to overcome obstacles? Absolutely. But united this village can change the course that we are on and figure out how to overcome these obstacles and sustain quality science, technology, engineering, and mathematics programs, our future. After all, we all want the same things: teachers and students, this nation's future, with the knowledge and skills to compete in this ever-changing world.

We are here today to turn ideas into action. . . . We need to take our science education system to the next level, to the level that will help us to be responsive to our future needs.

—Jacqueline Dorrance

2

The National Context

Key Points

- The goal of establishing high national standards often has been mistakenly interpreted as requiring standardization, but standardization ignores the differing needs of students, schools, and districts.
- Ideally, the curriculum drives the development of assessments, but today large-scale assessments often dictate the content of the curriculum and approaches to instruction.
- Teachers need high-quality professional development to use effective curricula and assessments to full advantage.
- Avoiding educational failure requires recognizing the factors in the early grades that influence later student success.
- Linking education in technology, engineering, and mathematics to science education, thereby creating a truly integrated science, technology, engineering, and mathematics (STEM) education, could have major implications for K-12 education.

In 1981 the newspaper *Education Week* published its first issue, two years before the National Commission on Excellence in Education released its report *A Nation at Risk*. In the April 22, 2009, issue, founding editor Ronald Wolk critically examined five basic assertions to show why the United States is still a "nation at risk." At the convocation, Kathy DiRanna, who began running an elementary science program in 1983 for

15

the Orange County Department of Education, used Wolk's assertions to place science education in California in a national context (Wolk, 2009).[1] Wolk's first assertion is

The best way to improve student performance and close achievement gaps is to establish rigorous content standards and a core curriculum for all schools—preferably on a national basis.

This is a fine idea, said DiRanna. "It would be unconscionable that we would have schools that would have anything less than high expectations for all students." But it is an idea "whose promise has not been actualized." More than two decades after the standards movement began schools are not near to making the changes needed to realize this goal. They have not received enough support to implement good ideas. And no one knows a methodology that is guaranteed to enable all students to master rigorous content standards.

"The objective of high standards was translated into standardization— the idea that schools would look the same and offer uniform programs. But such a goal is impossible to achieve and counterproductive," DiRanna said. Teachers inevitably have different knowledge and beliefs about how student learning takes place. They may also have mandates from their schools or districts to take different approaches to instruction. For example, some districts may require **direct instruction**[2] rather than **inquiry-based learning.**

In addition, the K-12 population of students in California is extremely diverse and becoming more so. For example, as noted in Chapter 1, one-quarter of California's students are English language learners. How quickly students acquire language depends partly on experiences they have outside school, and teachers have little control over such experiences. Yet these students are supposed to meet the same national and local educational goals as all other students. Individual schools and programs need to be different to meet the needs of a diverse group of English language learners, DiRanna said. "Something that is uniform cannot work. We have to sit down and think about how to redesign the way we do school in order to think about addressing these kinds of issues."

The differences among students also extend to their out-of-school experiences. For example, Dennis Bartels of the Exploratorium observed that the informal science system in the United States is the most robust of any in the world and may account for the relatively high scores of U.S.

---

[1]To view this presentation, see http://www.nasonline.org/site/DocServer/DiRanna-Reviewing_the_past_Looking_Ahead.pdf?docID=54983.

[2]Boldface words in this chapter are defined in a glossary at the end of the chapter.

fourth graders on international comparisons of scientific proficiency. "The after-school community . . . is actually driving a lot of kids who never get any science in schools into thinking about scientific careers," he said.[3]

Wolk's second assertion is

Standardized test scores are an accurate measure of student learning and should be used to determine promotion and graduation.

This assertion corresponds with the accepted view that the curriculum should drive instruction, instruction should drive assessments, and information from those assessments should lead to further revision and improvement of the curriculum. Another way to think about this triad of factors, according to DiRanna, is that content standards should drive **formative assessments**, formative assessments should lead to **summative assessments**, and these summative assessments should inform the revision of content standards.

However, the lines of causation in today's education system are essentially reversed, said DiRanna. Standards drive the large-scale assessments, and those in turn drive formative assessments and instruction. "We see this time and again in districts that are taking the California Standards Test," said DiRanna. The creators of the tests are "cutting and pasting the released items, calling them valid and reliable benchmarks, and giving them to school districts to use in the classroom. . . . It is entirely a backwards design."

As a result, teachers instruct their students on how to fill in test answers, read sentence stems, and choose the best answer. They "teach [students] tricks to do the test as opposed to looking at learning," DiRanna said.

The effect of this backward approach on teachers' attitudes has been dramatic, according to DiRanna. For example, in a survey done by the Teachers Network (2007), more than 40 percent of teachers thought that the No Child Left Behind legislation encouraged rote drill. Only 3 percent thought that it encouraged them to improve their teaching, while 44 percent thought that it made them eliminate curriculum material that was not on tests. And 69 percent agreed that it contributed to teacher burnout and to teachers leaving the field.

Ideally, standards and research on the understanding of students guide the development of formative assessments, which in turn drive

---

[3]A major report released just prior to the convocation (National Research Council, 2009) provides evidence that many kinds of out-of-school experiences can encourage interest and learning in science.

summative assessments.[4] Teachers need to use the idea of **learning trajectories** to understand where students are in their thinking. Research on the understanding of students needs to be used to design assessments for the classroom that also can contribute to summative tests.

Wolk's third assertion is

We need to put highly qualified teachers in every classroom to ensure educational excellence.

No one can argue with that goal, said DiRanna, but the requirement of the No Child Left Behind legislation to achieve this goal by 2006 "didn't quite happen." Placing highly qualified teachers in every classroom is as difficult as raising student understanding, for several reasons. Bright young people are not being attracted to education. (Even the sons and daughters of teachers tend to say, "I don't want to work like you," DiRanna observed.) Teaching is a complex act that is difficult to learn through existing teacher preparation programs. Once teachers begin teaching, they often encounter harsh conditions, such as decaying buildings, inadequate laboratories, and professional isolation. New teachers can have a hard time being accepted by their more senior colleagues. And teachers still are not seen as professionals, either in California or across the nation.[5]

In the best possible world, DiRanna pointed out, teachers love their students and love the material they are teaching, so that students also acquire a love for that material. But those conditions do not always hold. Policies and practices therefore need to be in place to acknowledge and overcome hindrances to learning.

The effective use of assessments poses particular challenges in terms of professional development. In a recent effort supported by the Center for Assessment and Evaluation of Student Learning, teachers who had excellent teaching practices were monitored for their assessment practices (Herman et al., 2006). But their assessment practices were not nearly at the same level as their teaching. "Very few of them were really reflective on their practice in terms of what assessment is telling them and how they monitor and adjust their instruction." Teachers needed professional development on each of the aspects of assessment to help them think about what they want their students to know, how to determine if they know it, and what to do depending on whether they do or do not know it. After

---

[4]Additional information about the relationship between in-class assessments and large-scale, high-stakes tests is available in National Research Council (1999, 2003).

[5]For additional information about treating teaching as a profession, see National Research Council (2000a).

training to build a framework of assessment knowledge, teachers were much more sophisticated about their assessment practices. "Teachers are capable of doing this kind of work, [but] they need to be helped along the way."

For practicing teachers, professional development is essential but currently has many problems. It has to be done over time to be effective. It has to focus on teacher practice, not just on learning what the content happens to be. Research has begun to show that changes do take place when professional development is extended over time. One study indicated that 49 hours of professional development helped increase student achievement by 21 percentile points, DiRanna pointed out (Yoon et al., 2007). And 80 hours of professional development can change teacher practice; 160 or more hours can produce actual changes in the classroom.

Wolk's fourth assertion is

The United States should require all students to take algebra in the eighth grade and higher order mathematics in high school in order to increase the number of scientists and engineers in this country and thus make the United States more competitive in the global economy.

This goal is part of a movement toward mandated learning, DiRanna observed. For example, California is now mandating that every student should study algebra by the eighth grade. But most students who do well in algebra in the eighth grade already have a knowledge and ways of thinking about mathematics that are distinctive in the fourth and fifth grades. In contrast, most students who will have trouble with algebra in the eighth grade already have fallen behind in understanding in the fourth and fifth grades. "We have to think not about the eighth grade but go back and say, 'What does K-8 look like so that children can be successful in the eighth grade?'"[6]

Wolk's final assertion is

The student dropout rate can be reduced by ending social promotion, funding dropout prevention programs, and raising the mandatory attendance age.

However, research has shown that students do not drop out of high school on a whim, DiRanna observed. They have been thinking about it for a long time. Already in elementary school they may have tuned out what school has to offer them. As with mandated learning, attitudes and

---

[6]For an additional perspective, see National Research Council (1998).

motivation need to be examined in the earlier grades to understand why students drop out of high school.[7]

> *We have to think not about the eighth grade but go back and say, "What does K-8 look like so that children can be successful in the eighth grade?"*
>
> —Kathy DiRanna

Reflection on these assertions makes for a rather gloomy view of education, but there is also a bright side, said DiRanna. First, the past 25 years have produced many new research findings from the cognitive sciences and education. For example, researchers have shown that metacognition—the ability to reflect on what one knows and does not know—begins very early in life.[8] As another example (which was pointed out by Dennis Bartels of the Exploratorium), research in the field of mathematics has shown that having particular skills in elementary school is a very good indicator of numeracy in later years.[9] Those skills could be built into curricula, assessments, and teacher practice in elementary school to encourage success in later classes.[10]

The importance of conceptual understanding also is much better understood now than it was 25 years ago. The slogan DiRanna applies in her work is "What's the big idea?" California, in contrast, chose to emphasize many small details in its standards, she stated. As a result, the challenge in California is to place the existing standards within a larger story. In an age of information, students need to learn how to construct meaning, not just repeat facts. They also need to be able to apply knowledge to novel situations. This requires a rich curriculum that encourages students to do this kind of work.[11]

Finally, research is becoming available that demonstrates the efficacy of inquiry-based instruction (National Research Council, 2000a; Yoon et

---

[7]Additional information about research on high school dropouts is available from National Research Council (2001a) and National Research Council and Institute of Medicine (2003).

[8]For additional information, see National Research Council (2000b).

[9]For an additional perspective, see National Research Council (2001b). A similar report also has been published that focuses on learning science in grades K-8 (National Research Council, 2007a).

[10]A report for teachers of science in grades K-8 (National Research Council, 2007b) has been published that is based on National Research Council (2007a).

[11]For additional information, see papers and presentations from a National Research Council workshop, "Exploring the Intersection of Science Education and the Development of 21st Century Skills," held February 5-6, 2009 (http://www7.nationalacademies.org/bose/21st_Century_Skills_Workshop_Homepage.html).

al., 2007). This approach to learning can help close the achievement gap, encourage reasoning and evidence, and lead to higher postinstruction test scores.

FROM SCIENCE EDUCATION TO STEM EDUCATION?

Several speakers at the convocation described the many benefits that would be gained by linking science education more strongly to education in technology, engineering, and mathematics, creating a truly integrated STEM education, and many participants continued to discuss and reflect on the implications of this approach to teaching and learning throughout the convocation. Greg Pearson, a senior program officer with the National Academy of Engineering (NAE), devoted the most time to the subject.[12] He previewed some of the conclusions of an NAE study that seeks to understand, capture, and analyze activities related to teaching engineering to K-12 students.[13] The goal of the study, Pearson said, "is to provide guidance to key stakeholders regarding the creation of K-12 engineering curricula and instruction practices that focus especially on the connections of the STEM subjects."

Engineering is a problem-solving process. One useful way to describe engineering, according to Pearson, is that it is "design under constraint." The laws of nature studied by science are one such constraint. But there are many other constraints, including money, time, materials, human resources, regulations, values, ethics, and even politics.

Since the early 1990s, an estimated 5 million K-12 students have engaged in classroom engineering education, the study has found. An estimated 10,000 teachers have had some sort of professional development related to teaching engineering, almost all of it connected to a specific curriculum. Many such curricula have been developed, and the NAE study report analyzes 16 in detail.

One finding is that many curricula struggle to include mathematics in meaningful ways. Science is more prevalent and is often presented as a way to uncover laws of nature or aspects of science that can be used in engineering design. Technology is often used to provide context for engineering design or to show how engineering can be applied.

Very little teacher professional development is occurring in this area. Also, there are no comprehensive content standards or frameworks for engineering education, as there are for science, mathematics, and tech-

---

[12]For access to this presentation, see http://www.nasonline.org/site/DocServer/Pearson_-_NAE_study_on_K-12_engineering_education.pdf?docID=54990.

[13]For additional information about this study, see National Academy of Engineering and National Research Council (2009).

nology. Science and engineering "are still largely treated separately both in teacher professional development and in the curriculum," Pearson said.

The NAE report encourages people to think about ways that the STEM subjects can become more truly interconnected through an engineering framework. Pearson laid out three possible scenarios of ways in which this could happen. In the first scenario, which he described as the status quo, K-12 engineering education remains largely below the radar of educators, policy makers, and the public, and national K-12 STEM reform continues to focus on mathematics and science. In an innovation model, the number and size of K-12 engineering programs grow dramatically, so that they become more visible and popular with educators and policy makers. In the third scenario, which Pearson called the paradigm shift, engineering facilitates a shift in STEM education toward "true" interconnection, and significant change and disruption of the status quo occur as the nation adopts a new vision of STEM education. This scenario would have major implications for how teachers are educated, the development of curricula and assessments, and many other aspects of STEM education.

Ethan Lipton, a participant from California State University at Los Angeles, urged those at the convocation to embrace Pearson's third scenario. Science education in California used to be 20 years ahead of science education elsewhere in the country. Now it is 20 years behind, Lipton said. California could regain its leadership by "re-envisioning" science education in the context of technology, engineering, and mathematics. Such a change could have a major effect on the choices students make, the skills of the workforce, and the development of the economy. "I encourage you, when you think of STEM, to think of T and E," Lipton said. "Don't think about fighting over the pie. Think about what we need for our students and how we can work together to do that."

> *I encourage you, when you think of STEM, to think of T and E. Don't think about fighting over the pie. Think about what we need for our students and how we can work together to do that.*
> —Ethan Lipton

Glossary of Education Terms

Direct instruction—instruction in which a teacher explicitly teaches a set of skills or knowledge base through lectures or demonstrations

Formative assessment—assessment that takes place during the learning or teaching process to gauge student progress and improve learning

Inquiry-based learning—a student-centered means of education in which students seek answers to challenging questions with guidance from teachers and other resources

Learning trajectory—the conceptual and cognitive path by which learning might proceed

Summative assessment—assessment after the completion of learning activities to gauge the mastery of concepts or development of skills

3

Science Education in Action

Key Point

- A demonstration of effective science teaching with a diverse group of fifth graders and a poster session and demonstration of scientific concepts by sixth graders showed how engaging and informative science education can be.

Because it was unclear whether all participants had actually personally experienced or observed active, student-engaged science teaching and learning in the early grades, the organizing group decided to model this kind of teaching and learning early during the convocation. Thus, on the first morning, attendees were able to participate in a fifth grade science class taught by Nancy Chung, a teacher at Hicks Elementary School in the city of Irvine (see Figure 3-1). Her students joined the attendees at their tables in the Beckman Center to help the adults through the lesson. This lesson, along with the "Poster Session on Science Investigations" that is described below, engendered considerable subsequent discussion among participants throughout the convocation about the success of these approaches to teaching and learning, the critical role of teachers in making these kinds of lessons accessible and interesting to students, and how science education in California might be changed if such approaches were to be adopted statewide.

FIGURE 3-1 Fifth grade teacher Nancy Chung leading a hands-on science lesson for her students and convocation participants. Photo courtesy of Sue Neuen.

This is a lesson that had the potential to get messy, Chung said, but her attitude was, "Why not? Students need to have fun in the classroom. That's my motto. Keep the students engaged and get their hands dirty, because that's when they learn the most."

> Keep the students engaged and get their hands dirty, because that's when they learn the most.
>
> —Nancy Chung

The objectives of the lesson were to identify the state of matter of an unknown substance called "Oobleck," understand what scientists do when faced with an unknown substance, and discover that Oobleck is a "non-Newtonian fluid" that has the properties of both a liquid and a solid.[1] To the theme music of Mission Impossible, she announced to her students, "We're here today on a very special mission. We have been asked to identify a mysterious substance that I'm holding in my hand right now. This mysterious substance has so many scientists completely baffled. We do not have an official name for it yet, but for now they're calling it 'the Oobleck,' named after the mysterious precipitation from Dr. Seuss's book. It is our duty today as young scientists to contribute our input from our tests and our observations. Are you ready for this challenge?"

Chung began with a quick review of the characteristics of solids, liquids, and gases. At her school, students use a Smartboard, an inter-active learning tool, to explore the properties of molecules in a solid, liquid, and gaseous state. "Do you remember that, boys and girls?" Chung asked. "Who can tell me what the molecules were doing in a solid state?" According to one student, they were vibrating very slightly but not moving fluidly. "How about at a liquid state? When the water was at room temperature, what were the molecules doing?" One student said that they still kind of stuck together but could move around, and another pointed out that a liquid takes the shape of a container. "What about a gas? You remember moving the [Smartboard] temperature up on the thermometer, and what would happen? The water would start to boil and you would see the water vapors coming up. What was happening to the molecules?" One student said that the molecules were moving very fast and going off the screen, and another said that a gas was very compressible. "Whereas the solid—can you compress that?" Chung asked. Not really, the students replied.

---

[1] Additional information about this classroom activity is available at http://lawrencehallofscience.org/gems/GEM200.html.

Chung held up a green substance in a plastic container. "We're going to look at this substance." First she had the students guess whether the substance was a solid, liquid, or gas, and each put his or her name on a Post-it note and placed the note on charts labeled with those headings. The students then put plastic tablecloths on their tables. "To find out whether it is a solid, liquid, or gas, we're going to run a series of tests," said Chung. Each group had a sheet and a pencil to record the results of the tests, which were described on handouts that accompanied the Oobleck.

"What would you like to find out [about this substance]?" Chung asked. Students replied: What is Oobleck made of? Where is it from? How thick is it? "Do you mean its consistency?" Chung asked. "Yes," the student said. "I have a question for you," Chung said. "Is Oobleck a solid, liquid, or gas?" Then she invited students to write down any additional questions they wanted to answer with their tests, along with whether it is a solid, liquid, or gas.

Chung next asked the class what happens after a testable question is written down. A student responded that the next step is to generate a hypothesis. "Can you tell us what a hypothesis is?" asked Chung. The student responded that it is a prediction about what you think will happen. "Is it any kind of random prediction?" asked Chung. No, the students said, it is an educated prediction with reasons why you think something will happen.

The students then began doing a number of tests, including the pour test, the poke test, the squeeze test, the sink test, the roll test, and tests they devised on their own. Chung led them through the initial poke test. Carefully, they used chopsticks to stab the Oobleck without knocking over their containers. "Was that weird?" Chung asked. "I thought it was liquid. Shouldn't it go all the way through? What did it feel like? Somebody raise their hand and tell me what it felt like."

"Like there was some kind of substance at the bottom," one student said. Others said that the chopstick went through the top layer very easily and kind of got stuck toward the bottom.

She had the students write down the time and date, since "scientists always write down when they are doing an experiment, what day it is and what time it is." She then had them slowly press their chopsticks into the Oobleck, demonstrating that the chopsticks could pass slowly through the substance. They wrote down what they did and what they observed. They then wrote down what they thought the Oobleck was based on that test. "And if you think it's more than one thing, go ahead and circle that one," Chung said.

While the students began performing additional tests on the Oobleck, Chung described a technique she uses to maintain the focus on specific learning objectives. Classrooms in her district have signs bearing the acronym SWBAT, which stands for "students will be able to." Because of the presence of the word "bat" in the acronym, Chung occasionally sings the Batman theme song and then asks her students to read the lesson objective in unison. She demonstrated, and the students read: "To classify the unknown substance as a solid, liquid, or gas, using prior knowledge of the characteristics of the states of matter."

After about 10 minutes, Chung said, "I'm so sorry to interrupt your having so much fun. I'm going to ask you a question. Raise your hand if you would like to change your answer. If you originally said, 'I think Oobleck is a liquid,' raise your hand if you think it might be something else. Lots of hands. Now raise your hand if you're thinking, 'I'm going to stick to my answer.' Nobody? Okay. Discuss with your tablemates what you think your conclusion is. Do you think it's a solid? Is it a liquid? Is it a gas? Or is it more than one thing?"

"Please write down what you think the Oobleck is. And did you know that scientists use observations to collect data, but must agree on the accuracy of their data with other scientists? Then flip your papers to the very back." On a Venn diagram on the last page of their handouts, the students wrote the word Oobleck where they thought it should go, including in areas of the graph that include more than one category. "If you think it's all three, then write Oobleck in the middle, where all three overlap." Then she directed their attention to the words "Something to Think About" at the bottom of the page.

"Does anyone know what Oobleck is made of? Do you think I went to a super fancy science store and said, 'Please give me the Oobleck ingredients'?" But several students already knew that Oobleck was a mixture of corn starch, water, and green food coloring. Chung said that they were right, and that the corn starch and water were mixed in about a two to one ratio. "I'm thinking of ratio. What if I did the two to one ratio but instead of corn starch to water I did water to corn starch? What would happen?"

"It would be more liquidy," one student said. Another said that it would not be as hard when it was poked. A third said it would flow faster when poured.

"These are all very good answers. Maybe you can try that at home."

The second question under "Something to Think About" was "Why do these simple ingredients form such a mysterious substance?"

"Because you have two different things that have been combined," one student said.

"What was the previous unit that we covered that talked about something like that?" Chung asked.

"The mixtures and solutions unit," a student replied.

Chung said: "We talked about the difference between a physical reaction and a chemical reaction. What do you think this is? Do you think it's the result of a chemical reaction or a physical reaction?" A chemical reaction, one of the students answered.

> *Sometimes I tell you I know everything, but of course I don't, because I'm a lifelong learner, too.*
>
> —Nancy Chung

"Raise your hand if you think it's a chemical reaction," Chung said. The majority of students raised their hands. "Now raise your hand if you think it's a physical reaction. Hmm. How can you tell? Is there something you can do to test that, besides look it up on the Internet or ask a teacher? If it's a physical reaction, can you somehow separate it and get it back into its original form? So I wonder if I could get this back to its original form. Nicholas, what do you think?"

"You could evaporate the water," her student said, leaving just the corn starch and the food coloring.

That would work for a physical reaction, Chung said. "I don't know if you could separate the corn starch and the food coloring," she added. "There's probably a way, but I don't know. But what do I tell you all the time? As a teacher, do I know everything? No, I don't know everything. Sometimes I tell you I know everything, but of course I don't, because I'm a lifelong learner, too. So what can we do if we're faced with a problem or a question that we don't know? Does the learning stop here? Or are you going to take it to the next level and find out why the corn starch and water combine in such a mysterious way? Or what causes corn starch to behave that way. How you can separate those? Are those questions that you can be thinking of? So that's where your line of learning, your 'LOL,' comes in."

> *So what can we do if we're faced with a problem or a question that we don't know? Does the learning stop here?*
>
> —Nancy Chung

"We're running out of time. But there are more questions, like are there other substances that have similar characteristics to the Oobleck? I heard someone say quicksand. So the next time you're stuck in quicksand, what can you do if you want to get out? So brainstorm with the adults at your table about some substances with properties like those of the Oobleck. Come up with different industries where substances like Oobleck can come into play, whether it's in the medical field, the construction field, transportation, communication, agriculture, energy, packaging, defense, or space exploration. So go ahead and talk about that and we'll clean up afterwards."

"Thank you. I hope you had a lot of fun playing—I mean learning—with the Oobleck."

Excerpts from the California Fifth Grade Science Standards[a]

1-a Students know that during chemical reactions the atoms in the reactants rearrange to form products with different properties.

1-f Students know differences in chemical and physical properties of substances are used to separate mixtures and identify compounds.

1-g Students know the properties of solid, liquid, and gaseous substances.

Investigation and Experimentation

Scientific progress is made by asking meaningful questions and conducting careful investigations. As a basis for understanding this concept and addressing the content in the other three strands, students should develop their own questions and perform investigations. Students will

a. Classify objects (e.g., rocks, plants, leaves) in accordance with appropriate criteria.

b. Develop a testable question.

c. Plan and conduct a simple investigation based on a student-developed question and write instructions others can follow to carry out the procedure.

d. Identify the dependent and controlled variables in an investigation.

e. Identify a single independent variable in a scientific investigation and explain how this variable can be used to collect information to answer a question about the results of the experiment.

f. Select appropriate tools (e.g., thermometers, meter sticks, balances, and graduated cylinders) and make quantitative observations.

g. Record data by using appropriate graphic representations (including charts, graphs, and labeled diagrams) and make inferences based on those data.

h. Draw conclusions from scientific evidence and indicate whether further information is needed to support a specific conclusion.

i. Write a report of an investigation that includes conducting tests, collecting data or examining evidence, and drawing conclusions.

---

[a]The California Science Standards for Grades K-12 are available at http://www.cde.ca.gov/BE/ST/SS/documents/sciencestnd.pdf.

POSTER SESSION ON SCIENCE INVESTIGATIONS

Following the science lesson, the convocation broke for a poster session featuring the fourth through sixth grade students at two local elementary schools, La Veta Elementary and Nohl Canyon Elementary (see Figure 3-2). The questions investigated by the students were

- Which objects will be attracted to the charged balloon?
- How can the dry mixture be separated?
- What happens when the chemicals are combined?
- What can you tell about the age of the tree by observing the rings?
- How are topographic maps made?
- How does the angle of the ramp affect the speed of the car?
- Can you detect the magnetic field in each box?
- What will happen if one light bulb is removed from the series circuit and the parallel circuit?
- What will happen when a flood occurs where there is an existing water source on a slope?

FIGURE 3-2 Students demonstrate their work in science to convocation participants. Photo courtesy of Maureen Allen.

Convocation attendees walked from poster to poster as teams of students explained how they conducted their lessons and what they learned. At one of the posters, Megan, a sixth grade student from La Veta Elementary School, demonstrated the acceleration of a car down a ramp. She and her two partners had used a light gate to measure the time a toy car spent rolling down a ramp given different inclinations of the ramp. They then calculated the car's speed according to the formula $s = d \times t$, where s equals speed, d is distance, and t is time. As the angle of the ramp increased, the speed of the car did so, too.

"Hands-on science goes in your mind better," Megan said. "If it's just in a textbook, it's harder to absorb."

Hands-on science goes in your mind better. If it's just in a textbook, it's harder to absorb.

—Megan, sixth grader

4

Exemplary Programs

Key Points

- Exemplary programs in California and elsewhere in the nation, several of which were described at the convocation, demonstrate that highly effective science education not only can be implemented but also has many benefits.
- The Beckman@Science Program in Orange County has provided more than 1 million students with hands-on, inquiry-based science classes.
- The Merck Institute for Science Education has improved the teaching and learning of science through an emphasis on student performance and participation, instructional practice, school culture, and district policies.
- The Leadership and Assistance for Science Education Reform Program in Washington state has brought together the stakeholders involved in science education to pursue a multifaceted agenda of improvements.
- The Woodrow Wilson National Fellowship Foundation has sought to strengthen science education through fellowships to undergraduate science majors intending to become teachers.

Throughout the convocation, attendees learned about K-8 science education programs that have met with notable success. Presentations on Beckman@Science, the Merck Institute for Science Education, the Leadership and Assistance for Science Education Reform (LASER) Program in Washington state, and the Woodrow Wilson National

Fellowship Foundation demonstrated the potential for innovative initiatives to achieve the goals called for at the convocation.

BECKMAN@SCIENCE[1]

In 1997, at the age of 97, chemist, industrialist, and philanthropist Arnold O. Beckman stated that he wanted to improve the way science was taught to young children because of his concern that the United States was losing its competitive edge in science. Beckman "wasn't extremely fond of the textbook-only method of teaching science," said Jacqueline Dorrance, the executive director of the Arnold and Mabel Beckman Foundation. He remembered vividly childhood experiments in his makeshift laboratory and felt that, by providing teachers with the tools they needed, students could have similar experiences.

In cooperation with the community, California State University at Fullerton, the National Science Resources Center in Washington, DC,[2] the Discovery Science Center in Santa Ana, California,[3] the Exploratorium in San Francisco,[4] and the Ocean Institute in Dana Point, California,[5] the Beckman Foundation created the Beckman@Science Program. The program fosters hands-on, inquiry-based science for students in kindergarten through sixth grade and emphasizes quality curriculum, professional development, inquiry-based materials, and community and administrative support.

The program was offered to all of the schools and districts in Orange County for the first five years after the program was instituted in 1998. At the end of that period, 6 private schools and 15 school districts in Orange County had accepted the offer (see Figure 4-1). The foundation chose to work in Orange County for several reasons. First, that is where the foundation is headquartered, which enabled progress to be monitored more closely and in person. Second, Orange County is the fifth largest county in the nation, with more residents than 22 states, which means that the potential impact of the program could be enormous. Finally, Orange County has a very diverse population, with a higher proportion of English language learners than the state average. "We felt that if we were successful in Orange County, others might be encouraged by our

---

[1]PowerPoint slides from this presentation are available at http://www.nasonline.org/site/DocServer/Dorrance_Panel_Presentation_-_final.pdf?docID=54995. Additional information about Beckman@Science can be found at http://www.beckman-foundation.com/@Science/prog_info.htm.

[2]Additional information is available at http://www.nsrconline.org.

[3]Additional information is available at http://www.discoverycube.org.

[4]Additional information is available at http://exploratorium.edu.

[5]Additional information is available at http://www.ocean-institute.org.

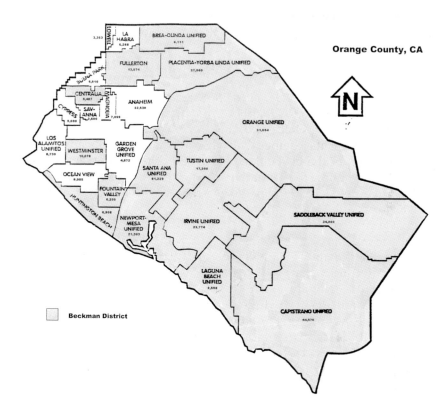

FIGURE 4-1 A map of Orange County, showing the school districts that are part of the Beckman@Science initiative.
SOURCE: Arnold and Mabel Beckman Foundation.

success and we could serve as a model for those interested in improving their science education programs," said Dorrance.

Since the program was begun, tens of thousands of teachers, principals, and administrators have taken part in Beckman@Science activities. They have been able to see firsthand the "the wonder, accomplishment, and excitement in the students' expressions as they participated in hands-on science lessons," said Dorrance.

The program has worked hard to provide districts with a solid foundation to support sustainability of the initiative. Professional development has been a critical component of the program, reflecting the direct correlation between success in the classroom and the quality and quantity of professional development that teachers receive (Yoon et al., 2007). Since the program began, approximately 24,000 teachers have received

23,544 Teachers Trained

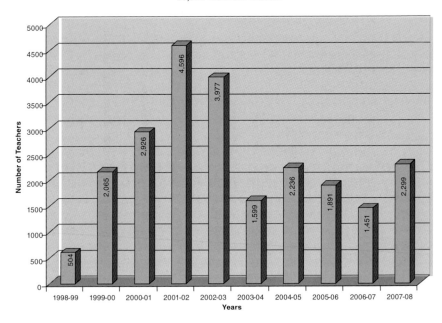

FIGURE 4-2 Beckman@Science has provided more than 23,000 teacher professional development sessions during the program's first decade.
SOURCE: Arnold and Mabel Beckman Foundation.

professional development to promote hands-on, inquiry-based science in their classrooms (see Figure 4-2). These 24,000 teachers have, in turn, delivered more than 1 million hands-on, inquiry-based science lessons since Beckman@Science began (see Figure 4-3).

During the first five years of the program, the Beckman Materials Science Center offered professional development and material support to every public and private school. More than 11,000 teachers have participated in the training offered at the Materials Science Center. These teachers were prepared to use more than 27 kit titles that were offered for grades K-6.[6] In partnership with nearby colleges and universities, the program offered teachers courses on science content and on helping students prepare and maintain science notebooks. In addition, the program offered kit specialist and train-the-trainer courses, all geared to provide teachers with the highest quality professional development possible.

---

[6]For information about the specific materials used in the program, see http://www.beckman-foundation.com/@Science/prog_info.htm.

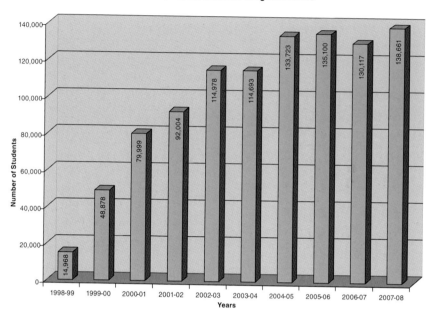

FIGURE 4-3 More than 1 million inquiry-based kits have been used in Beckman@Science schools.
SOURCE: Arnold and Mabel Beckman Foundation.

Following their work at the center, the teachers were able to borrow the kits for up to 12 weeks. The program trained approximately 150 teacher leaders, and 600 teachers received rigorous preparation in their districts in advanced inquiry strategies, which they could then disseminate among the other teachers in their districts. The Materials Science Center and individual districts also have hosted community science awareness nights, which have given thousands of people from all walks of life opportunities to learn about and experience hands-on, inquiry-based science. The teachers, principals, administrators, students, and parents who have been inspired through these outreach efforts will help sustain the program in the long run, Dorrance observed.

The Beckman Foundation offered each participating district up to 12 years of funding to be used exclusively for professional development, materials, and administrative and community support. The program also holds monthly meetings of coordinators for the districts, and these meetings have led to "lasting friendships, admiration, respect, and trust," said Dorrance. The coordinators discuss challenges, share successes, and exchange strategies for implementing, improving, and sustaining hands-on science instruction in their districts.

In addition, principals' breakfast meetings are held in the Beckman districts in which the principal and a parent organizer from each of the schools are invited to learn about high-quality, hands-on, inquiry-based science, share in the successes of their students and teachers, and discuss the challenges that their schools are facing. They are encouraged to continue to provide the leadership support necessary by providing teachers with adequate time and opportunities for professional development.

The districts "have learned their lesson well," Dorrance said. All of the districts have established materials refurbishment systems that provide teachers with hands-on, inquiry-based materials aligned to meet the California science standards. The purchasing, scheduling, delivery, and refurbishment of these materials have required substantial effort on the part of the districts. Without these systems in place, the program would probably fail, since teachers could not be expected to supply such materials on their own. The majority of the districts are using a minimum of two different science kits per grade level per year, with many using three. In 2008, a group of districts formed the Kids@Science Foundation,[7] a not-for-profit organization dedicated to working with schools, districts, and communities to collaboratively promote, support, and enhance hands-on, inquiry-based science education for children. "The Beckman districts are serious about sustaining inquiry-based science education," Dorrance said.

To assess its effectiveness and make changes, the program has gathered data from a variety of sources, including an external evaluator, WestEd, district annual reports and strategic plans, site visits, test scores, and conversations with teachers, students, parents, principals, and administrators. These evaluations have shown that teachers in the Beckman districts report changes in the way they teach science. Teachers have indicated that they have become more effective at teaching science, that their knowledge of scientific concepts has increased, and that they have become much more confident in teaching science.

There also has been a notable difference in the depth of science being taught. For example, middle schools report that students exposed to Beckman@Science have a broader and deeper science foundation than before their exposure. Middle schools also say that they see an increase in students involved in extracurricular activities linked to science.[8] In

---

[7]A website is under construction but was not available at the time that this report was prepared.

[8]A recently published report (National Research Council, 2009) provides an extensive review of science learning in informal environments.

addition, high school teachers report an increased number of students requesting science classes in participating districts.

Teachers also have reported positive changes in their students related to science. They say that their students have found science more enjoyable, interesting, and fun. They also say that students grasped and applied scientific concepts more easily. Students have become better observers, critical thinkers, communicators, and collaborative workers. The districts report that the quality of the science notebooks created by the students has increased and that the use of these notebooks reinforces the mathematics and literacy skills of students as well as improving their understanding of science. Parents report that science is now a topic of conversation at many dinner tables, according to Dorrance.

Finally, California standardized test scores show that since 2005—the first year that California schools were tested in science—the Beckman@Science districts have outperformed the non-Beckman districts at both the county and state levels. The 2008 test scores reveal that 87 percent of Beckman districts scored at the proficient level or above compared with 30 percent of non-Beckman districts (see Figure 4-4). At the state level, 93 percent of the Beckman districts scored proficient or above compared with 50 percent of the non-Beckman districts.

The Beckman@Science districts have faced and will continue to face challenges, Dorrance pointed out. The greatest challenge today is the ongoing state budget crisis, which is likely to result in layoffs of key administrators and teachers. Class sizes will increase and some teachers will change grade levels, which will require that they be given additional professional development in the kits for their new grades. New and less experienced teachers and administrators will be hired and will require extensive training. And while this is occurring, the demand for this dynamic and innovative science program will continue to grow. "Districts will need to work together, as they have done in the past, to find creative solutions to these problems," said Dorrance.

> *Districts will need to work together as they have done in the past to find creative solutions to these problems.*
>
> —Jacqueline Dorrance

Following the presentations about Beckman@Science and the other exemplary programs described below, breakout groups explored the lessons to be learned from the programs. Reporting for one of two breakout groups that discussed Beckman@Science, Angelo Collins observed that such programs may require explicit exit strategies for participating

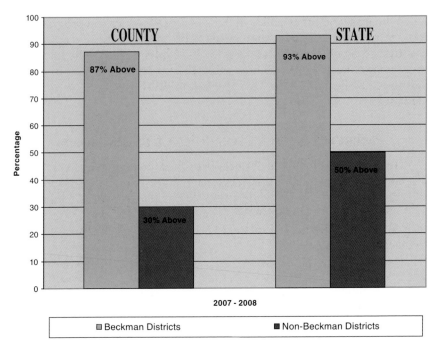

FIGURE 4-4 Districts participating in Beckman@Science scored at the proficient level and above at a much higher rate than for the country or state as a whole. SOURCE: Arnold and Mabel Beckman Foundation.

schools and districts, since no grant lasts forever. No-cost extensions of current groups could help establish some of the "feathering down" processes that are needed. In addition, the breakout group suggested that programs like Beckman@Science need to maintain a focus on innovations that were not in the original plan. These innovations might include responding to needs that have emerged since the original plan was developed, incorporating tools or knowledge that have become available since the original plan was prepared, or responding in new ways to knowledge gained as the original plan was implemented.

Ideally, the funder-funded relationship focuses on the values embodied in a program and not on money, Collins observed. Whether seeking funding, developing plans, evaluating activities, or reporting results, it is important for funders and program managers to highlight the outcomes sought and achieved by the program. And the creation and support of collaborations are essential if programs are to enact meaningful change.

Reporting for the second breakout group, Frank Frisch indicated that his group looked critically at the extent of involvement of politicians, school boards, and other policy makers in science education programs. In some respects, science education needs to be "sacrosanct," he said, so that it is treated as an essential element of K-12 education. Public advocacy for science education, perhaps by celebrities or other well-known public figures, could reinforce the message that science education is critical to the nation's future.

MERCK INSTITUTE FOR SCIENCE EDUCATION[9]

The Merck Institute for Science Education was launched in 1993 as a separate not-for-profit organization by Roy Vagelos, who was chief executive officer. "The institute's mission," said its executive director, Carlo Parravano, "has been to improve the teaching and learning of science, with the ultimate goal being to improve student performance and engagement in science."

The institute began its work by creating partnerships with school districts in the states of Massachusetts, New Jersey, and Pennsylvania, where Merck has a major presence. It originally focused on grades K-8; more recently it also has begun to work with high school teachers. Most of its funding has come from the Merck Company Foundation, with two additional grants from the National Science Foundation (NSF) on local systemic change and on math/science partnerships.

The institute has devoted a significant amount of effort to program evaluation. "We want to accumulate knowledge," said Parravano. "We think about program evaluation as an integral part of every program that we design, and we also make a very strong effort to have all of the stakeholders involved in evaluating the program."

> We think about program evaluation as an integral part of every program that we design.
>
> —Carlo Parravano

An important tool in planning for program evaluation is a theory of action—a set of ideas underlying a program that can be tested, elaborated, and refined. Developing such a theory requires consensus on the desired results. It also can establish a relationship among the activities

---

[9]PowerPoint slides from this presentation are available at http://www.nasonline.org/site/DocServer/Parravano_-_Beckman_Center_Presentation__April_29-30__20.pdf?docID=54985. Additional information about the Merck Institute for Science Education can be found at http://mise.org/mise/index.jsp.

being planned and among the resources brought to bear on a project. And it helps organize and focus evaluations, because it brings to the surface what some of the research questions should be and what sources of data are available.

Throughout its history, the institute has measured changes in four areas: student performance and participation, instructional practice, school culture, and district policies. These four areas have been examined using multiple sources of data, including standardized tests, enrollments, classroom observations, surveys, interviews, and case studies. Evaluations by such organizations as the Consortium for Policy Research in Education[10] and Westat[11] have included annual reports, formative evaluations, and four capstone reports that are on the institute's website.

One of the institute's prominent achievements, according to these reports, has been the creation of a model of professional development for science education, which some school districts are now extending into mathematics and language arts. This model has produced high levels of teacher participation, high rates of change in instructional practice, widespread use of standards-based instructional materials, dramatic changes in student experiences, and, as Parravano put it, "unfortunately, modest changes on standardized tests." Despite the test results, science remains a priority in the school districts, higher expectations have been established for all students, teacher expertise is highly valued, and district policies and strategies in areas like incentives and hiring have changed.

In addition to evaluation, sustainability has been a key component of the institute's efforts. (Chapter 5 discusses issues of program sustainability in detail.) According to Parravano, it is important for discussions of sustainability to take place throughout the life cycle of a program, from its early planning stages, to its modification, to its conclusion. In the past two years, for example, Merck's formal commitment to the institute and an NSF grant were coming to an end. That transition forced a serious discussion about what aspects of its efforts the institute wanted to sustain—the focus on science, a particular approach to professional development, an emphasis on building capacity in schools, collaboration among schools and districts, or all of this.

Through a review of the literature, the institute identified leadership and capacity as important factors in sustainability. It sought to answer the questions: Are there leaders in the districts who are knowledgeable about science education? Do they have a vision for science education? Can they articulate that vision? Are their policies aligned to support that vision?

---

[10]Additional information is available at http://cpre.org.
[11]Additional information is available at http://westat.com.

Another important factor that affects sustainability, the institute found, is teacher engagement and efficacy. Do teachers have the requisite knowledge and skills? Are there ongoing opportunities for professional development?

In addition, national and state policy environments are very important. Are assessments aligned with standards? Is there a commitment to reform among school principals and teachers? Do parents want to focus on improvements to science education? Are students engaged? Is student achievement improving? "The answers to these questions are critical," according to Parravano.

The institute has worked with districts to ensure the sustainability of science education programs by building leadership and capacity among teachers and principals. For example, as part of the Leader Teacher Institute, teams of teachers from school buildings committed to three years of professional development—three weeks each summer and sessions during the academic year as well. The program produced facilitators in professional development: coaches, mentors, principals, curriculum reviewers, and so on who are now capable of sustaining reform. "If a new superintendent comes in and says, 'Why are we focusing on science education to such an extent?' these leader teachers are there to demand the same kind of attention to science as before."

Engaging teachers is also critical. The professional development programs supported by the institute have been anchored in research while also relating directly to what teachers do in the classroom. The programs address in an integrated way curriculum, pedagogy, and learning, with plenty of attention to student needs and respect for participants' time, expertise, and experience. The programs combine intensive offsite learning with on-the-job professional development. As Parravano observed, "It's not either/or that's important, but a combination of both."

The institute also has tried to participate in discussions of statewide standards on curricula and professional development and has disseminated research findings on teaching and learning from the National Academies. That work "has paid off enormously," according to Parravano.

Merck has been supporting the institute's work for more than 15 years, and the company has been "incredibly patient" as the institute has learned what works best. The company also has encouraged the institute to be focused, both geographically and on K-8 science education, which has added coherence to its work in districts and statewide.

As with the Beckman@Science Program, the institute has faced challenges. The lack of quality assessments has been the greatest problem, according to Parravano. Assessments need to speak to teachers, to students, and to the public, "and we really don't have any of that right now." Turnover of teachers and administrators, which requires the continual preparation of new participants in the program, also has been an issue. And, finally, Merck recently purchased another company, Schering-Plough, which means that the company will grow from 60,000 employees to more than 100,000. "There are definitely going to be significant changes," Parravano said. "One of the challenges that we have is to make sure that Merck continues to value this kind of work."

Reporting from the breakout group that discussed the Merck Institute for Science Education, Greg Pearson detailed some of the many objectives that need to guide the development of better assessments. Different audiences need different kinds of data. Assessments need to demonstrate causal relationships whenever possible, but they also need to measure such indicators as "Are students asking better questions?" State requirements can greatly affect assessments—for example by narrowing definitions of performance to short-term student achievement gains—but the institute and businesses can push for more meaningful tests. A long-term "improvement infrastructure" is needed at the district level to maintain the focus on better assessments.

Pearson also reported on a breakout group discussion of whether to direct resources to expanding programs that have had demonstrated benefits. For example, the California Science Project[12] is a small program that has had a major impact. But state policy leaders can overlook small programs that district and local leaders would identify as effective. Furthermore, little infrastructure exists that would allow resources and innovations to be imported into a district or school.

Finally, Pearson's breakout group noted the pressing need for a network model of system change and sustainability, as opposed to a linear or strictly hierarchical model of change. Other reform efforts in science, technology, engineering, and mathematics (STEM), such as those in Ohio and Texas,[13] depend critically on networks. With networks, it is not possible for a single stakeholder to stop progress because there is not a single gatekeeper and because of the interdependence of the stakeholders in the network.

---

[12]Additional information is available at http://csmp.ucop.edu/csp/index.php.

[13]These efforts are being supported by the Bill and Melinda Gates Foundation. Additional information about the initiative in Ohio is available at http://www.battelle.org/spotlight/1-30-08stem.aspx. Information about the initiative in Texas is available at http://poly.rpi.edu/article_view.php3?view=6648&part=1.

WASHINGTON STATE LASER[14]

The Leadership and Assistance for Science Education Reform (LASER) Program in Washington state is a public–private partnership that has brought together the full range of stakeholders in science education, according to its codirector Jeffrey Estes. Schools decide whether to participate in the program, which is a project of the National Science Resources Center, an entity created by the National Academies and the Smithsonian Institution. Washington state is one of eight LASER sites that were selected as part of a 1999 NSF implementation and dissemination project. It is committed to a blueprint for improvement based on standards, shared goals, classroom-tested curricula, professional learning experiences for teachers, data-driven decisions, material support, administrative and community support, and a "living" improvement plan.

Results produced in part by the LASER Program, as measured by scores on standardized tests, vary greatly from district to district and from school to school, Estes observed. Nooksack, a small rural district near the Canadian border, has posted district-wide scores in science that are nearly 30 percentage points higher than the state average since becoming part of the program. But even in that district, the scores of the three elementary schools have increased at markedly different rates. "Although we're trying to move things along across the state, and we're trying to move things along as a district, we end up moving this along building by building," said Estes.

The LASER Program has learned to build on the strengths of individual schools and districts. For example, the Kennewick School District in southeastern Washington is well known for its attention to reading. This turned out to be an impediment to promoting science education, because district officials feared that an emphasis on science would detract from the focus on reading. However, the LASER Program helped district and school leaders think about how science might be an ideal partner for raising scores in reading. According to Estes, "You have to find what works in the school district."

> *You have to find what works in the school district.*
>
> —Jeffrey Estes

---

[14]PowerPoint slides from this presentation are available at http://www.nasonline.org/site/DocServer/Estes_WA_State_LASER_-_NAS_Convocation.pdf?docID=54984. Additional information about Washington State LASER Program can be found at http://www.wastatelaser.org.

The LASER Program adheres to a theory of action that involves knowledge of research, a shared vision, a supportive school and district infrastructure, and a focus on improving the teaching of science. Key individuals in leadership roles have proven to be critical in implementing this theory of action. These individuals have come not just from the education community but from the business community, higher education, and government.

"We view ourselves as trying to catalyze sustainable improvements by providing an opportunity for schools that's too good to refuse," Estes said. The program strives to ensure that initiatives are based on current research, that they are visible and well understood in the state, that they are supported by a healthy mix of funding, and that leadership comes from multiple sectors. The program creates challenging goals for schools. "We keep pushing. We keep nurturing. We keep trying to help schools progress and translate those goals into day-to-day practices that make a difference for kids."

The program relies on a regional support system that is augmented by statewide assistance. It also involves a diverse set of partners from business, education, the nonprofit sector, and local communities. A major focus is the provision of products and services that amplify the effect of investments in science education. In addition, the program is involved in such areas as curriculum and assessment development, professional development, and community support.

As one way to assess the effectiveness of these efforts, the LASER Program has sought to understand how the amount and quality of professional development has improved student learning. Using both end-of-unit assessments and statewide comprehensive assessments, the program has sought to relate results at the fifth and eighth grade levels to teacher professional development. The results of these assessments in turn have affected the design of the program.

Changes in the program are linked with its sustainability. With the program in its second decade, its leaders are trying to figure out how the program can evolve to keep pace with changes occurring in the state. A particular tension has been between science literacy for all and the need to fill the innovation needs of the state in the short term by producing university graduates in STEM fields. In resolving this tension, the program has had to pay attention to both tangible factors, such as instructional materials, professional development, and funding, and intangible factors, such as achieving a critical mass of activities and support, perceptions, and adaptation. "The key," said Estes, "is to maintain the program's core beliefs and values." (These issues are discussed in greater detail in the next chapter.)

The LASER Program also faces challenges. State standards have been revised and new assessments instituted. The state has a new superintendent of public instruction, and the State Board of Education has assumed an increased role in science. Difficult economic circumstances are affecting the mix of public and private resources. And more work is needed to change public perceptions to embrace science as a "new basic." The program will seek to overcome these challenges as it has in the past, said Estes—through intensive partnerships, reliance on a diverse set of stakeholders and funders, and the cultivation of leadership in key positions. "We look forward to the challenge," said Estes.

Reporting for the breakout group that discussed the Washington LASER Program, Rena Dorph listed some aspects of the program that are needed for sustainability. Fostering partnerships has been critical for the program's success, especially in its use of people who can act as bridges across organizations. Involving business in public-private partnerships has been important, because it offers to business a way of demonstrating the returns on investments. In addition, five elements of the program provide tangible outcomes that people can understand:

1. Continuing investment and dialogue among partners, even as the roster of partners changes.
2. Establishing and nurturing a "culture" of partnership.
3. Building on previous efforts that stakeholders recognize as successful.
4. Providing clear explanations to stakeholder groups, such as parents, about the importance of the initiative.
5. Developing strategies to deal with the inevitable transitions of key personnel.

The question of who leads the organization is especially important, Dorph noted, because LASER traditionally has been seen as outside the traditional power structure and as a neutral convener of stakeholders.

To ensure that professional development is ongoing and progressive, it is important to get a commitment from districts, especially those with resource constraints. It also is important to find people in the state who are highly qualified to offer professional development and to train new people to offer professional development over time.

For the program to be maintained, its mission needs to fit with missions of the lead organizations, so their support is a natural component of the program's work. In addition, those organizations need to have a passion for the program's mission and a sense of shared leadership and stewardship.

Finally, any such program will face difficulties. Among the ones identified by the breakout group are finding the leadership to hold the partnership together, dealing with the turnover of leadership at all levels, and empowering teachers to transfer what they learn from professional development to the classroom.

WOODROW WILSON FOUNDATION[15]

The Woodrow Wilson National Fellowship Foundation has been awarding fellowships to students to attend graduate school since the end of World War II. Since Arthur Levine became president of the foundation in 2006, it also has emphasized the preparation of mathematics and science teachers. Under a program Levine initiated, prospective teachers with an undergraduate degree in science receive $30,000 to attend a one-year master's program in teacher education. In return, they make a three-year commitment to teach in a high-needs school. They receive intensive mentoring to encourage them to remain in teaching. The foundation also has been working with universities to transform preservice teacher education in science and mathematics so that teachers are better prepared when they enter schools.

The program is focused on the state level because "statewide programs offer the most leverage," according to Levine. The foundation chose to work with Indiana, which is the 19th largest state in the country. The program produces 80 mathematics and science teachers per year. "That's not a lot," Levine observed. "But the simple reality is that, with those 80 teachers, we will increase the number of teachers who are certified annually by 25 percent." The foundation can have a much larger impact on a state like Indiana than on one like California, which hires approximately 2,000 science teachers each year. In addition, small states offer economies of scale in terms of recruitment, placement, and assessment.

A key element of the foundation's success in Indiana was the establishment of a statewide coalition that includes the highest echelons of state leadership to support mathematics and science education. In Indiana, the coalition consists of the governor, key members of the legislature, the chief state school officer, the state higher education executive officer, the university community, college and university presidents, faculty, school superintendents, school boards, unions, the business community, professional associations, and philanthropies. "That coalition was critical to us," said Levine. "What it meant was that when the governor didn't get reelected, the program didn't die. . . . If the chief state school officer left, or the state higher education executive officer left, we still have the same coalition. We just have new players in it. Programs can persist if you build the right coalition."

---

[15]Additional information about the Woodrow Wilson National Fellowship Program can be found at http://www.woodrow.org.

Programs can persist if you build the right coalition.
—Arthur Levine

The coalition in Indiana provided stability, joint responsibilities, and ownership of ideas, according to Levine. High-level representation on the coalition also provided a means of exerting pressure on parts of the science education system that need to change. For example, the governor could press for changes in the state university system through both formal and informal channels. "The sticks are as powerful as the carrots," Levine said. Funding from the foundation provided another incentive for change, with requirements for matching funds ensuring commitment on the part of educational institutions.

Reforming science education in elementary and middle schools will require many changes, Levine said. Universities must prepare more science teachers. Salary bonuses may be needed to recruit quality teachers to all schools. Professional development for practicing teachers needs to increase in quantity and quality. More facilities and equipment are needed.

One important step, said Levine, echoing advice from Clark Kerr when he headed the Carnegie Council on Policy Studies in Higher Education (1980), is "put a number on it." Instead of saying that new teachers are needed, say 300 new elementary school teachers are needed. Instead of saying that the curriculum must improve, say that a new fourth grade curriculum in the physical sciences is needed.

Many groups are focusing on better teacher education, including universities, school districts, not-for-profit organizations, such as Teach for America, state governments, the federal government, and for-profit organizations like the American College of Education. The Wilson Foundation has chosen to work largely with universities on teacher education, because universities still prepare the large majority of teachers. Universities also can sustain programs once those programs are established. And universities are the centers of scientific disciplines, and association with these disciplines can provide solid content for science teachers.

The foundation also has told universities that it wants to focus on outcomes, for both teachers and students. In addition, programs need to combine the arts and sciences and education, drawing strengths from different departments. The foundation wants third-party assessments of university programs and their outcomes. This clarity of expectations helps the universities and the foundation know what is important, in return for funding the fellowships and separate funding to reform university curricula. The existence of the coalition also encourages change, in that members of the coalition could both reward and punish universities for cooperating or resisting change.

The task is often difficult for universities, Levine said. For example, Ball State University, the largest producer of teachers in Indiana, broke its teacher education program into modules that it embedded in field experiences for future teachers. It created residencies for teachers to work on program design and residencies for professors in schools so they could spend time teaching in the schools in which they were preparing teachers to work. "We've told [the universities] that the whole world of science has changed. The whole world of math has changed. What we expect of schools is very different than we used to expect from them. We need a new coterie of science teachers and math teachers. What we want you to do is join us and let's invent that future together."

The foundation expected a memorandum of understanding signed by the president of the university, the school board, and the superintendent that included dates and deliverables. Funding for curriculum reform then hinges on meeting those dates and deliverables. For example, the universities "had 21 months to change their entire programs before any fellows arrived. If they hadn't changed the program, they got no fellows." In addition, vertical integration was emphasized. Recruitment, preparation, placement, retention, and professional development were tied into a single program instead of being reformed one by one.

"In the years ahead, we have the capacity to obtain the science education the children in our nation need," said Levine. "But achieving this goal requires moving with alacrity, establishing explicit roles for each stakeholder in the system, and instituting coalitions of the key actors in each state."

> *We have the capacity to obtain the science education the children in our nation need. But achieving this goal requires moving with alacrity, establishing explicit roles for each stakeholder in the system, and instituting coalitions of the key actors in each state.*
>
> —Arthur Levine

5

Fostering Sustainable Programs

Key Points

- Understanding how effective science education programs can be sustained requires an examination of the assertions and associated assumptions underlying those programs.
- Sustainability can be defined as the ability of a program to maintain core beliefs and values and use them to guide adaptations to internal and external changes and pressures over time.
- A comprehensive literature review has revealed more than 25 factors associated with the sustainability of effective science education for grades K-8, including some that have not been widely discussed before.
- Sustainability requires and expects that a program's operating principles are likely to be adapted to different circumstances as they are instituted in new places, but that its core beliefs and values will remain largely intact.
- Program planning should accommodate future as well as current goals.
- The critical components of effective programs need to be identified in clear language to learn from innovation.
- Patience, a long-term perspective, and flexibility are all critical to sustainability.

Several assertions underlie many efforts to improve science education programs, according to Jeanne Century, director of science education and research and evaluation at the University of Chicago's Center for Elementary Mathematics and Science Education.[1] Among the most prominent of these assertions are the following:

- Effective practices need to be identified.
- Those practices need to be scaled up and sustained.
- Decisions need to be based on evidence.

Embedded in each of these three assertions are important assumptions that need to be investigated to learn how effective science education programs can be sustained, Century said. Her research group at the University of Chicago has been investigating these assumptions through a comprehensive review of the literature on both effective science education as well as on sustaining reforms in economics, business, marketing, and health. This noneducation research offers "a different angle on the question" of sustainability, she said. "We are pretty insular in science education, and that hasn't served us well in research because we don't benefit from the work that other people have done."

> *We are pretty insular in science education, and that hasn't served us well in research because we don't benefit from the work that other people have done.*
>
> —Jeanne Century

PREVIOUS WORK ON SUSTAINABILITY

About a decade ago, Century was involved in a project that looked at nine school districts around the country that had sustained elementary science programs for between 10 and 30 years (Century and Levy, 2002). The study conducted surveys with teachers and principals, interviewed teachers and school district leaders, and analyzed documents and news clippings. One outcome of the project was case studies of the districts. But an even more interesting result was the identification of a number of factors extending across the districts that either contributed to or inhibited the sustainability of their programs.

A subset of the factors fell into a category that Century called the "usual suspects," because they are both important and often discussed.

---

[1]PowerPoint slides from this presentation are available at http://www.nasonline.org/site/DocServer/CenturyPresentation.pdf?docID=54982. Additional information about the University of Chicago's Center for Elementary Mathematics and Science Education is available at http://cemse.uchicago.edu.

These include instructional materials, leadership, accountability, money, professional development, policy alignment, and culture. "We know that we need these things to make a program happen," Century said. In addition, some factors emerged that Century categorized as unusual suspects, including adaptation, critical mass, perception, and quality. Whereas the usual suspects centered mostly on the implementation of programs, the unusual suspects were critical for sustainability. A third important product of the earlier study was a definition of sustainability. In the past, the term has had many different meanings. Based on the data collected for the study, the research team settled on the following definition.

> Sustainability is the ability of a program to maintain core beliefs and values and use them to guide adaptations to changes and pressures over time.

Thus, Century said, sustainability is not necessarily judged by the ability of a program to find additional funding or to be embedded in a district budget. Instead, sustainability is focused "on core beliefs, values, and adaptations."

FACTORS UNDERLYING SUSTAINABILITY

In their study, Century and her colleagues used a very broad array of search terms to identify papers connected in some way to the concept of sustainability. This process resulted in 30,000 abstracts. Team members read the abstracts and narrowed down the list to about 600 papers. They then coded the text in those 600 papers to identify the factors involved in sustainability. "We went through a very iterative process of clearly defining what every single factor meant."

The team first had to tackle the question of what is lasting in a program. Is it the program itself, the effects of the program, or the philosophy of the program? For example, in one case studied by Century and her colleagues, a program disappeared when funding was lost. But funding for the program was later restored and the program was resumed. Even during the period when the program was not operating, the philosophy of the program to rely on hands-on experiences and not on textbooks was retained.

Without determining exactly what about a program needs to be lasting, Century and her colleagues simply called that enduring quality "the it," where "it" may refer to the program, its impact, its philosophy, or some other entity. Similarly, they abandoned the word "sustainability" and referred to the quality of lasting as "the concept," with different aspects of this quality being labeled "concept A," "concept B," and so on. The researchers also emphasized the time frame over which change occurs, since different kinds of change can occur over different periods.

Century and her colleagues identified more than 25 factors that affect "the it." Among those factors are flexibility, adaptability, specificity, complicatedness, feasibility, and effectiveness. All of these are interconnected. For example, while effectiveness is an important characteristic of a program, it is only one factor. Similarly, when an effective practice becomes embedded in daily practice, it is sustained. But a practice can be embedded in daily practice and stay the same, or it can be embedded in daily practice and change.

Century and her colleagues identified several factors that emerged as especially important in their study. One factor encompasses the characteristics of people in an organization, including their experiences and points of view. In addition, the factor "elements of the internal environment," including the internal structures, social climate, and resources of a program, is important, as is the "external environment," which involves the political climate, students' opportunities for learning, and other external forces.

An unusual factor that the researchers identified was "emotional mediators," which include characteristics like trust, loyalty, and incentives. "These are things we don't usually hear people talk about when they're talking about the elements of reform. But we know that these things have a huge impact on why people do and don't do things."

"Fit" is a measure of whether "the it" is consonant with the values, beliefs, needs, and practices of the people involved in the program. For example, if the fit of a program is too close to current practice, then change is not really occurring, whereas if the fit is too far from current practice, people become so uncomfortable that they are not likely to change. "You need to find a sweet spot," she said. "If you find yourself in a place where you are very comfortable, it suggests that you're not pushing yourself enough." People change out of necessity and out of will. Feeling uncomfortable is a sign that you're doing something differently, and that's where the opportunities lie.

> *If you find yourself in a place where you are very comfortable, it suggests that you're not pushing yourself enough.*
> —Jeanne Century

The factor "mechanisms" has to do with the spread or scale-up of "the it." But does scale-up refer to the spread of a program, to the expansion of a program's benefits, or to the acceptance of a program's philosophy? "We didn't know what it meant, so we decided that we were just going to talk about types of movement, because we knew we had a grip on that. The 'it' moves from one place to another." Similarly, the idea of replication does not help much in understanding sustainability, which is more about translation and adaptation.

The interplay of the factors identified by Century and her colleagues is a dynamic and complex process. "All of these factors come into play at the same time, and they then all change, and then they all change again because the context and conditions around us are always changing."

REEXAMINING THE ASSERTIONS

Given the importance of these factors, the three assertions with which Century began can be viewed in a new light. The first assertion was that effective practices need to be identified. As an example, Century discussed the Slip! Slop! Slap! campaign in Australia (Montague et al., 2001). In an effort to encourage people to take steps to prevent skin cancer, the Australian government supported a program to encourage people to slip on a shirt, slop on sunscreen, and slap on a hat. These are easy things to do, said Century. But it took 10 years for the program to make much of a difference in people's behavior. "It's not that it wasn't effective. But it didn't fit with their belief systems, with current practice, or with the way people were functioning in their lives."

From a sustainability perspective, effective practices are useable, flexible, and resilient. "The it" is not necessarily the practice but something deeper.

The next assertion was that effective practices need to be scaled up, which in the most common formulation means replicating a program in multiple places. But as Century pointed out, teachers never teach a lesson exactly as it was written. They take their knowledge and expertise and adapt that lesson to their context and conditions, because that's how a lesson becomes effective. Any particular program, intervention, or model is going to change because of local circumstances. "It doesn't just transfer and replicate. It translates every time it moves. So why are we focused on replication when we really should be thinking about how to capture and understand the way things translate as they move from place to place?"

> *We really should be thinking about how to capture and understand the way things translate as they move from place to place.*
>
> —Jeanne Century

Translation requires more than the identification and replication of best practices. It requires hard thought about what is being replicated and how to replicate that quality. "It's not really about identifying the best practice," said Century. "It's about identifying the processes by which we can thoughtfully and then effectively help them move and last."

The process of moving a program requires consideration of the time frame. Program directors need to ask themselves whether they want a particular program to be exactly the same 5 years from now, 10 years from now, and 20 years from now. "Planning is not something that happens at the beginning," said Century. "It's something that happens all the time. [It's] an ongoing process by which we're constantly thinking about the factors, adjusting our program, and hopefully improving."

Identifying the critical or essential elements of a program is important in this planning process. Some components may be necessary at the beginning of a program but are not needed later. Some parts of a program may need to be discontinued while other parts remain. "This is a good thing, because you toss away the things that aren't necessary anymore."

The process of involving others in change through a collaborative process can be misleading if it involves convincing others to agree with and accept a preexisting model. From a sustainability perspective, the important thing is to have a process in place by which a model can be continuously examined and improved. A collaborative change process involves creating change together.

Knowledge does not exist to be discovered or revealed, said Century. It is fabricated in practice through talking, writing, or acting. Information dissemination is different from knowledge building. The process of change needs to be captured systematically and clearly, using a shared language, so that understanding about the process of change accumulates.

"We don't want to make changes that last," said Century. "We really want lasting change. We want continuous, ongoing change." Change needs to be accepted rather than resisted. A program may retain some of its essential elements as it moves from one place to another, but it will translate as it does so. That's how investments in science education will endure in the long term.

> *We don't want to make changes that last. We really want lasting change.*
> —Jeanne Century

Finally, Century's study has revealed that multiple sources of evidence, not just student outcomes, need to be examined to inform decisions about which elements of a program should last. The climate, the time frame, and different perspectives all need to be considered, because "everything counts."

CORE CONCEPTS

Two core concepts emerged from the study. First, "we need to think about what we're doing now at the same time that we think about where we're going." That's hard, because getting a program in place can be an all-consuming task, but it is necessary. "If we just focus on the program now, all of the context and conditions around us are going to keep changing. . . . It's not going to wait for us to get our program in place." This is a problem in science education, according to Century, because the goals for science education have not changed much since the launch of Sputnik more than a half-century ago. "We need to rethink what the goals for science education are now. And at the same time, we need to be thinking about what the goals are for 50 years from now so that we can be thinking about what we're doing now that's going to help us get to where we need to be."

The second core concept is the importance of learning from innovations. Existing science education programs have many things in common and some things that are different. By identifying the critical components of the programs in clear language, the similarities and differences among programs can be studied. Every investment in a program needs to have a return in knowledge gained about the program. This requires that researchers develop a shared language with which they can accumulate knowledge in the field. This is not typically done in science education today, according to Century.

One conclusion she has drawn from her analysis is that funders need to be patient and embrace mistakes. Fields, such as cancer research, have been funded for decades resulting in a steady accumulation of data and knowledge. It will take time to accumulate knowledge once a system to analyze education is established. In addition, funders, researchers, and practitioners need to be willing to look across programs to accumulate knowledge. And parts of the system need to be willing to accept failure, since some programs that are innovative will fail. But people learn from failure. "It's not about the perfect program. It's about the process," Century said. "We need to create an environment of learning, not an environment of success, . . . because that's really going to teach us what's going to work and what's not going to work."

Business, in particular, needs to bring a long-term perspective to its efforts in education. Century said that she was surprised recently to learn that business assigns individuals the task of thinking about where the business is going to be in 10 years. "Business can help bring us this perspective and, of course, participate with us, not just as a supporter, but as a real partner."

Finally, the leaders of reform initiatives need to consider flexible adaptation. Everyone involved in science education needs to be willing to give up aspects of a program that are not working. "We need to be continuously flexible and adaptable. That's how we're going to get sustainability."

Century and her colleagues are still in the middle of their project. Given the importance of engaging in a collaborative change process, they have decided to make all of their work public. They have put all their analysis into an open collaborative research environment called Researchers Without Borders[2] so that anyone can follow the research and join in on the effort. Besides trying to put into practice the lessons they are learning, posting their preliminary analysis allows them to be systematic and explicit about their work and form partnerships with program practitioners.

---

[2]For additional information, see http://www.researcherswithoutborders.org.

6

Rising to the Challenge

Key Points

- A statewide coalition dedicated to creating an outstanding science education system could address the problems facing K-8 science education in
  California today.
- Each sector represented in the coalition could play a distinct role while
  contributing to the coalition's overall goals.
- Professional development, the time devoted to science in K-8 classrooms,
  and the establishment of an infrastructure for ongoing educational improvement all require special emphasis.
- The time to act is now, while science education occupies a position of
  prominence on state and national agendas.

As noted in earlier chapters, the problems surrounding elementary and middle school science education in California are obvious and systemic, but many promising options exist to begin solving those problems. The question is how to make progress.

A CALIFORNIA SCIENCE EDUCATION COALITION

The approach that was discussed intensively at the convocation is the formation of a coalition dedicated to creating an outstanding science education system throughout California. The coalition would include all of the stakeholders involved in science education, including K-12 teachers and administrators, parents, education researchers and evaluators, colleges and universities, professional societies, government at all levels, research institutions, business and industry, informal science education organizations, and philanthropies. Such a coalition would provide an opportunity for all stakeholders to agree on a shared vision and common goals.

The California Council on Science and Technology (CCST) and the California State University System (with California Polytechnic State University as the lead institution) have received a planning grant from the Bill & Melinda Gates Foundation and the S.D. Bechtel, Jr. Foundation to produce a blueprint by the end of 2009 for an advocacy campaign on behalf of science education.[1] According to CCST's executive director Susan Hackwood, this ongoing activity has made it an ideal time to establish a science education coalition in California. The advocacy campaign will be "long-term, on target, with many people saying the same thing over and over." It will involve well-known public figures, chief executive officers (CEOs), industry associations, and others who are in a position to make a difference. CCST is planning two levels of advocacy. The "grass roots" are the schools, districts, parents, and local industries that have a stake in science education. The "grass tops" are the prominent leaders who can work directly with government, industry, and foundations to foment change. "We're going to rely on you for the content," Hackwood said to the convocation participants. "But what comes out of this meeting will get traction within our organization, and within other organizations, too. . . . We can help build the campaign to make it happen."

Marilyn Edling of California Polytechnic State University said that she was putting together an advisory board for the campaign, which will include "people who are very powerful at a national level and a state level who can really take this campaign to Sacramento." The California Teacher Advisory Council of the CCST also can offer guidance to the coalition.

Each partner in the coalition would have a specific role to play. These roles were discussed throughout the convocation and particularly during a breakout session on the second day, when attendees separated into groups organized by sector. Rapporteurs for each group were asked to highlight up to three points that their sectors of the "village" could contribute to future efforts. The following section summarizes the discussions

[1]PowerPoint slides from this presentation are available at http://www.nasonline.org/site/DocServer/Hackwood__CA_STEM_Innovation_Network.pdf?docID=54989. Additional information about the California STEM Innovation Initiative is available at http://ccst.us/publications/2009/2009STEM.pdf.

from each of these sectors as they were reported to all participants in the convocation's final session.[2]

Business and Industry

Business can contribute in many ways to the establishment and functioning of a science education coalition in California. For example, it can provide project management specialists to facilitate the strategic planning process with other stakeholders. It can produce the foundational materials for a science education campaign. It can identify individuals in the state who are most likely to effect changes and send charismatic CEOs and sales-oriented people to promote the vision for changing K-8 science education. More broadly, business and industry can work with education communities in California to develop and articulate a vision of the state's future that is founded in strong science, technology, engineering, and mathematics (STEM) education.

According to Matthew Gardner from BayBio, which represents about 500 member companies in California, surveys of the organization's CEOs show that "the number 1 issue . . . is science education." Business leaders do not know the best way to solve the problem, he said, "but if you tell them, they will run with you." Gardner also emphasized that the business and industry sector would use the ideas that emerge from this convocation and all subsequent activities to emphasize issues that are critical for elected leaders to consider, especially as they run for reelection in 2010. On behalf of his working group, Gardner emphasized the importance of articulating all such plans clearly and developing plans for action by January 2010, when state races for office begin in earnest.

Higher Education

There are 200,000 undergraduates in the University of California system, 400,000 undergraduates in the California State University system, and 2.5 million students in California community colleges. These students represent a tremendous opportunity to nurture future teachers and supporters of science education. If they could be engaged in inquiry, exploration, and discovery, they could experience the many benefits of authentic science.

Representatives of higher education in California talked about several specific proposals. The higher education community could commit to developing a STEM fellowship program to work with K-8 teachers and

---

[2]PowerPoint slides from this presentation are available at http://www.nasonline.org/site/DocServer/Final_reports_from_sectors_of_the_village.pdf?docID=55001.

students. Fellows could include undergraduates, teacher candidates, graduate students, postdoctoral fellows, and STEM professionals. Colleges and universities also could develop STEM institutes and courses to provide preservice education and professional development for K-8 teachers and school administrators. In particular, they could develop master's degree programs focused on K-8 STEM education to prepare education specialists in these areas for elementary and middle schools.

The master's degree program is especially important, according to Joan Bissell, director of teacher education and public school programs for the California State University system. "Today at the master's level, STEM experts are trained for high school. We need people who can be STEM specialists in the elementary grades, who can train other elementary teachers, and who themselves can offer special periods where we start to think about elementary schools being organized differently."

Education Research and Evaluation Community

The research and evaluation sector can gather and provide the evidence, along several dimensions, that will be required to make the case that K-8 science education is critically important. Science education at these levels provides a foundation for later learning and attitudes toward science. It also develops individual cognitive skills and capabilities that are vitally important to society.

In addition, the research and evaluation sector can help illuminate the current status of K-8 science education and indicate whether quality education at these grade levels is happening or not. For example, it can determine whether opportunities to learn science at these grade levels are adequate.

Finally, the research and evaluation sector can make the case that effective K-8 science education *can* happen. It can document successful programs across the country, produce videos demonstrating student learning and engagement, and provide evidence that successful science education can occur at the levels of individuals, schools, and states.

Foundations

The foundations that currently support STEM education are a small subset of those that fund K-12 education. Nevertheless, they occupy a pivotal position in strengthening science education. They have a unique convening role—for example, one idea developed at the convocation is that the foundations represented there could convene a broader group of foundations to elevate science education as a priority. They also can provide assistance in many areas of policy. Many foundations combine leverage in local communities with an ability to take projects to scale.

Foundations also have certain limits. They should be viewed as catalysts, not as systemic, long-term funders. They require specific proposals with clear outcomes, projected budgets (including the magnitude of projected costs), and time lines.

Informal Science Education Institutions

Informal institutions like science centers and museums can serve as neutral catalysts for innovation, assemble partnerships for increased resources, and act as hosts for regional science resource centers, which in turn can provide learning opportunities for teacher and students.

Informal institutions also can leverage a large sphere of influence in a political campaign. In particular, the boards of directors and other stakeholders in informal science institutions can be extremely influential.

National Organizations

Federal agencies and national organizations can have a significant effect on K-8 science education in California—for better or worse. For example, Bruce Alberts pointed out that several federal agencies "produce curriculum materials for their own specialties, having no idea what an elementary teacher's life is like and how impossible it is for them to deal with all these things. Many of those resources could be much better directed toward some common vision of more fundamental science for everyone."

Several representatives of national organizations expressed their eagerness to contribute to the coalition's efforts. For example, Alan McCormack from San Diego State University, who recently has become president of the National Science Teachers Association (NSTA),[3] was previously president of the Council for Elementary Science International,[4] the affiliate of NSTA representing elementary science education. He said that he plans to increase the emphasis on elementary science education in the NSTA. "The theme of my presidency will be to promote elementary science," he said. "I'll do everything I can to make things better."

Dean Davis, a Boeing scientist and engineer and precollege deputy director for educational outreach for the American Institute of Aeronautics and Astronautics (AIAA),[5] said that "the AIAA offers, in LA, for example,

---

[3]Additional information is available at http://nsta.org.

[4]Additional information is available at http://www.cesiscience.org/.

[5]Additional information is available at http://aiaa.org/.

1,500 scientists and engineers that can come to your schools . . . and provide you with professionals who can describe how we use science in the
real world." AIAA also offers scholarships to educators and activities for
students.

Other Sectors

The village envisioned at this convocation also involves other critical
sectors, and these sectors were discussed throughout the convocation by
a number of presenters and as part of the general discussions.

Parents and other members of local communities can be both targets
of and participants in a California science education coalition. They need
to be convinced that learning science is important for *their* children, not
just for children in general. First, science is an excellent preparation for
a very wide variety of careers. As Susan Pritchard of the California Science Teachers Association[6] said, "when parents think science, they think
medical doctors, they think pharmacists, they think nursing or health
care." They do not think about the many jobs that require a technical
background and may be closely tied to a student's interests. "We need to
let our students know at an early age what kinds of career opportunities
are out there."

Science is the best way to teach children how to detect a bad claim—
whether a political claim, a commercial claim, or a scientific claim—said
Dennis Bartels of the Exploratorium. It is a message that resonates with
parents. "If you go up to a kid's parent and you say, 'Listen, do you want
your kid to be really easily lied to or manipulated or fooled?' Most parents
would say, 'Hell no. Don't let my kid do that.' Well, then, that's why they
need to take science, not because they're going to be an engineer. They
may be and that's great, but that's not really the reason for every kid from
K to 12 to learn science. It's to equip them with the antidote to the next
charlatan so that they can make good decisions."

> *Science is the best way to teach children how to detect a bad claim—*
> *whether a political claim, a commercial claim, or a scientific claim.*
> —Dennis Bartels

John McDonald of Stone's Throw Communications emphasized the
difficulty of making "people outside of this room care about these things."
An effective message has to be developed about why science education

---

[6]For additional information, see http://www.cascience.org/csta/csta.asp.

is important. "If you don't find ways to make the public care, then the legislature is not going to care, and you're not going to make the kind of progress you need." Similarly, Mark St. John of Inverness Research emphasized the importance of building a public campaign to make the public aware of the problems and potential for science education. "Creating public demand may be a very important part of the strategy."

> *If you don't find ways to make the public care, then the legislature is not going to care, and you're not going to make the kind of progress you need.*
>
> —John McDonald

Several speakers also pointed to the importance of individuals or institutions that can act as champions for specific science education initiatives. "Most success stories, almost to the letter, have at least one champion of the project," said Rena Dorph of the Lawrence Hall of Science.[7] These champions can work on many levels. They may help teachers access materials and professional development, support inquiry, marshal external funding to support science, provide instructional leadership to teachers, or keep districts and schools focused on science. They are "the keepers of the passion," said Dorph. "They spark, they incite, they create a buzz for continuing to do science in the face of these very constraining circumstances."

Bartels observed that many successful programs have a third-party institution that can sustain the program, whether a university, national laboratory, science museum, corporation, or some other institution. Representatives of these organizations can assure new leaders in a school district that they are attending to science. The response of leaders, Bartels said, is, "Great, I can worry about reading, or I can worry about something else."

Susan Neuen of the California Science Center[8] observed that a coalition for science education in California will need a staff first for it to be established and then to be effective. "Until there are people who are actually working on the establishment of a coalition full time, it's not going to happen," she said. In addition, the coalition needs to be part of the National Alliance of State Science and Mathematics Coalitions (NASSMC),[9] which is an umbrella organization for state coalitions of business, education, and public policy leaders united for systemic change in STEM education for all students.

---

[7]For additional information, see http://www.lhs.berkeley.edu.

[8]Additional information is available at http://www.californiasciencecenter.org.

[9]Additional information is available at http://www.nassmc.org.

ELEMENTS OF A COALITION AND POINTS OF EMPHASIS

Dennis Bartels laid out a specific four-point agenda that could be pursued by the coalition:

1. Create 12 regional science resource centers spread across the state funded at $2.5 million each. Every school in every part of the state would have access to these centers. They would help with professional development and curriculum implementation as well as provide materials to support inquiry-based science education.
2. Devote $2,500 per teacher for staff development in science, with the funds going to the districts to support the use of the new curricula. This proposal has made some headway in the California legislature in the past, Bartels noted, but has run up against resource constraints.
3. Establish several state leadership centers funded at several million dollars each focused on problems specific to science education. Such problems might include integrating science with language arts, science for English language learners, or science policy at the state level.
4. Join with other groups in California and across the nation to work for common standards in science. The primary reason for such a change, according to Bartels, is that California "can't afford to develop our own assessment system anymore. If we buy into a common set of standards, . . . the money goes to creating authentic performance assessments and other kinds of much more appropriate assessments behind those standards."[10]

In 4 to 6 years, such a plan could make a huge difference, said Bartels. It could guide investments in California while coordinating with national efforts to improve science education. Bartels estimated that his proposal would cost $40 million a year, a very small fraction of the total expenditures on education in the state.

POINTS OF EMPHASIS

Many speakers at the convocation pointed to specific issues affecting K-8 science education that need to be pursued, both inside the coalition and outside it.

---

[10]A month after the convocation, the National Governors Association and the Council of Chief State School Officers announced that 49 states and territories have joined an initiative to develop a set of core standards in English language arts and mathematics that states can voluntarily adopt. It is expected that common standards and assessments for science will follow the publication of standards in these other two subject areas. For additional information, see National Governors Association (2009) and http://nga.org.

Professional Development

Bartels observed that research has demonstrated the importance of the changes that occur in teaching after teachers receive at least 40 hours of professional development (Pasley, 2002; Simpson and Banilower, 2004). "When it goes from 40 to 80 or to 120 hours, suddenly you're on a real steep part of the S-curve, especially when you're working with really good curricular materials." Professional development needs to be fashioned around the curriculum that teachers are expected to teach, especially at the elementary level, since teachers have to rely on the curriculum materials that are available to them. This level of professional development costs somewhere between $2,500 and $4,000 per teacher, Bartels estimated (Noyce, 2006). That money probably has to be managed at the school district and state levels, Bartels added, with school systems and teachers "voting with their feet" for the types of services and infrastructure that meet their needs.

Jerry Valadez, director of the National Science Education Leadership Association,[11] pointed out that many inquiry-based materials are unused because teachers have not received professional development in how to use them. In addition, many teachers are not supported or encouraged to lead the kinds of after-school and extracurricular activities that can interest students in science. As a result, fewer children are participating in activities like science fairs or science Olympiads.

Bissell talked about the importance of creating a seamless boundary between preservice teacher education and inservice professional development. Some good programs along these lines have existed in the past, but the approach needs to become much more widespread.[12] The preparation of middle school science teachers will undergo a major change in the next few years, she said, because of a new general science credential that recently took effect in California.[13] This change represents a good opportunity to improve the teaching of science and prepare teachers to use science education to overcome some of the learning difficulties of students, according to Bissell.

California's colleges and universities will play a central role in this professional education, which emphasizes the importance of developing effective higher education teacher preparation and professional development programs. Anne Marie Bergen, the chair of the California Teacher Advisory Council,[14] noted the need to blend pedagogy and content in

---

[11]Additional information is available at http://www.nsela.org.

[12]A report from the National Research Council (2000b) discusses in detail the concept of seamlessness between preservice and inservice education.

[13]Additional information is available at http://www.cascience.org/csta/leg_credentialing.asp.

[14]For additional information, see http://ccst.us/ccstinfo/caltac.php.

teacher preparation, so that instructional strategies are woven into the content that future and current teachers are learning. "If I am the learner and I can see from the learner's point of view, it is extremely beneficial for me." Nancy Chung of the Tustin Unified School District pointed out how valuable ongoing professional development is for teachers. "It's not just that you're trained once and you're done. Having newer training and learning new methods are also very important, so you're not so outdated, since there are so many things coming up."

Legislative Initiatives

In some cases, specific pieces of legislation or other mandates could lead to substantial improvements in science education. For example, several people at the convocation thought that minimum instructional times for science from kindergarten through eighth grade need to be mandated. One breakout group discussed a minimum of 90 minutes per week for grades K-3, a minimum of 120 minutes per week for grades 4-5, and a minimum of 225 minutes per week for grades 6-8. This amount of preparation would prepare all students to take the equivalent of at least four years of science in high school.

Standardized testing in science can lead to greater time being devoted to the subject in K-8 education, but the scope of the testing can be critical. For example, California now tests its fifth graders in science, but the test covers just the fourth and fifth grade science standards. "If we had a test for all the elementary grade levels, what kind of impact would that have?" said Pritchard. "I think it would be very positive. We would have more materials, equipment, and teaching time in the younger grades." Similarly, the eighth grade assessment in science does not cover the sixth and seventh grade standards, which has the effect of de-emphasizing science in those grades. In general, said Pritchard, greater balance among the large subject areas of the curriculum in both assessments and instruction could help boost the prominence of science in K-8 education.

Legislation can address other issues as well. Bonnie Brunkhorst of California State University, San Bernardino, made the point that the Chinese government has legislated that teachers in China receive inquiry-based training. "They're beginning to prepare their people with inquiry education in science. It's time that we have some legislation to do the same here."

Legislative initiatives will require that science education have a greater visibility in the political process. John Kenny of PASCO Scientific noted that a science education coalition could increase the salience of the subject with legislators, especially if groups of people go to the capitol to lobby and speak with legislators and the governor. "My frustration has been that when there is a science bill being considered at the Capitol, nobody shows up. . . . We have to . . . bring out 50 individuals to any hearing that ever happens at the Capitol. That's the strength of a coalition."

Being in contact with the handful of people who exert disproportionate control over a particular issue is essential, according to Bartels. That observation reinforces the value of involving people who have good relations and connections with legislators and government officials. Science is a rare topic on which both political parties can agree, Bartels said. The challenge is to make science education a prominent issue among the many other problems facing the state. Along the same lines, said Gerald Solomon of the Samueli Foundation,[15] it would be advisable to start working with the candidates running for governor of California before the 2010 election, so that the groundwork is laid for shaping policy after the election.

Margaret Gaston, president of the Center for the Future of Teaching and Learning,[16] noted that the positions advocated by the coalition need to be based on "sound and reliable data," so that the coalition can give legislators sound advice. Dean Gilbert from the Los Angeles County Office of Education proposed that a clearinghouse be established that could draw links between proposed legislation and the effects on education. And Nancy Taylor of the San Diego County Office of Education said that a clearinghouse could include information about the history and successes of past programs. "We need to tell the California story so that we can build this cohesive coalition and move forward with these central ideas."

Legislators have an incentive to become involved in science education because of the way the issue permeates other issues. Susan Hackwood pointed out that virtually every study done by CCST calls for improvements in STEM education. "There's not a single issue that comes across our table to work on that doesn't involve STEM education," she said. "STEM education is at the foundation of all the major ideas that are emerging that will drive our economy. I'm thinking of the energy and climate programs that we have on the state's agenda, or the health care information technology, or the stem cell initiatives. They all involve knowledge of science."

> *STEM education is at the foundation of all the major ideas that are emerging that will drive our economy.*
>
> —Susan Hackwood

---

[15]Additional information is available at http://www.samueli.org.
[16]Additional information is available at http://cftl.org.

An Infrastructure for Improvement

Mark St. John of Inverness Research[17] emphasized the idea of establishing an infrastructure for educational improvements. Just as the Golden Gate Bridge has a crew of employees who work on maintaining the bridge full time, the improvement of schooling needs to be ongoing and continuous. "It's not something that can be done episodically."

Three levels of funding are necessary, said St. John. The most immediate level is the funding needed to schools to operate, while the most distant is the funding that foundations and others can provide for short-term, catalytic innovations. What is missing is a middle level of funding for continual improvement. "Steady-state funding for ongoing improvement efforts is the crucial missing link," he said.

This improvement process, several speakers noted, includes meetings like this convocation that bring together the stakeholders in science education. Philip Smith of the Space Grant Education and Enterprise Institute[18] urged the conveners of such meetings to "broaden the circle" as much as possible to bring in everyone interested in science education. Meetings provide an opportunity for stakeholders to settle on a unified vision of the future. And if a particular stakeholder is not included, the solutions they propose may not be what others might want.

TAKING ADVANTAGE OF THE CURRENT OPPORTUNITY

Recognition of the importance of the elementary and middle school years has placed a new emphasis on science education in those grades. "K-8 education is the place where we sow the seeds," said Jay Labov of the National Academy of Sciences. "This is where kids become interested, either to go on in careers in science, mathematics, engineering, or technology or to become the informed citizens that we will require for the workforce and for making the kinds of decisions that democracy requires." The challenge is to prepare all of California's children, "everybody," said Labov, "regardless of where they live, regardless of their socioeconomic background, race, ethnicity, or gender. We need to be offering these opportunities to everybody, and it begins in K-8 education."

During his after-dinner talk at the convocation, Arthur Levine described the unprecedented opportunity that currently exists. "Science education is hot," he said. Over the past-quarter century, many issues have moved to the top of the educational priority list, including curricula, class size, small schools, governance, charters, vouchers, school boards,

---

[17]Additonal information is available at http://www.inverness-research.org/index.html.
[18]Additional information is available at http://www.sgeei.org.

leadership, teachers, in-training development, preservice preparation, early childhood education, and so on. "This is the moment for science education. We're at the top of the list right at this moment, and California is the place for it to occur given the wealth of intellectual resources and the extraordinary need for excellence in science education."

But the moment will pass quickly, Levine warned. During the 2000 presidential election, education was near the top of the concerns registered in public opinion polls. During the 2004 election, it was number 5. In the 2008 election, it did not make the list. The economy, war, terrorism, health care, and energy had all passed education in the public's estimation.

Demographics are working against a continued focus on education. Baby boomers wanted good schools for their children, but many of their children are now completing high school or older. Instead, baby boomers are focusing on quality health care for their parents, and soon for themselves.

Maintaining momentum behind reform requires that science education be tied to national priorities, said Levine. The case needs to be made that education is important "because health care is important, because energy is important, because the economy is important, because defense is important."

"Not since Sputnik have the prospects for improving science education been more promising," Levine concluded. "The leadership of this effort in California is sitting in this room right now. . . . The future of this state depends on you."

References[*]

Bardhan, A., and Kroll, C.A. (2003). *The New Wave of Outsourcing.* Research Report Series No. 1103. Berkeley, CA: Fisher Center for Real Estate and Urban Economics. Available: http://repositories.cdlib.org/iber/fcreue/reports/1103/.

Bayer Corporation. (1995). *Teachers Feel Unqualified and Unprepared to Teach Science.* Press release, April 19.

California Budget Project. (2008). *Planning for California's Future: The State's Population Is Growing, Aging, and Becoming More Diverse.* Sacramento: Author. Available: http://www.unitedwayla.org/getinformed/rr/research/demo/Pages/Page1072.aspx.

California Council on Science and Technology. (2002). *Critical Path Analysis of California's S&T Education System.* Riverside: Author. Available: http://ccst.us/publications/2007/2007TCPA.php.

California State University Center for Teacher Quality. (2008). *Systemwide Evaluation of Teacher Education Programs in the California State University, 2007.* Long Beach: Office of the Chancellor, California State University.

Carnegie Council on Policy Studies in Higher Education. (1980). *The Carnegie Council on Policy Studies in Higher Education: A Summary of Reports and Recommendations.* San Francisco: Jossey-Bass.

Century, J.R., and Levy, A.J. (2002). Sustaining your reform. Five lessons from research. *Benchmarks: The Quarterly Review of the National Clearinghouse for Comprehensive School Reform, 3*(3), 1-7. Available: http://sustainability2003.terc.edu/media/data/media_000000000208.pdf.

Dorph, R., Goldstein, D., Lee, S., Lepori, K., Schneider, S., and Venkatesan, S. (2007). *The Status of Science Education in the Bay Area: Research Brief.* Berkeley: Lawrence Hall of Science, University of California. Available: http://lawrencehallofscience.org/rea/bayareastudy/pdf/final_to_print_research_brief.pdf.

---

[*]All URLS listed were active as of June 28, 2009.

Herman, J.L., Osmundson, E., Ayala, C., Schneider, S., and Timms, M. (2006). *The Nature and Impact of Teachers' Formative Assessment Practices.* Los Angeles: National Center for Research on Evaluation, Standards, and Student Testing. Available: http://www.cse.ucla.edu/products/reports/R703.pdf.

Marshall, R., and Tucker, M. (1992). *Thinking for a Living: Work, Skills, and the Future of the American Economy.* New York: Basic Books.

Montague, M., Borland, R., and Sinclair, C. (2001). Slip! Slop! Slap! and SunSmart, 1980-2000: Skin cancer control and 20 years of population-based campaigning. *Health Education & Behavior, 28,* 290-305. Available: http://heb.sagepub.com/cgi/content/abstract/28/3/290.

National Academy of Engineering and National Research Council. (2009). *Engineering in K-12 Education: Understanding the Status and Improving the Prospects.* Committee on K-12 Engineering Education, L. Katehi, G. Pearson, and M. Feder, Eds. Washington, DC: The National Academies Press. Available: http:// www.nap.edu/catalog.php?record_id=12635.

National Research Council. (1996). *National Science Education Standards.* National Committee on Science Education Standards and Assessment. Washington, DC: National Academy Press. Available: http://www.nap.edu/catalog.php?record_id=4962.

National Research Council. (1998). *The Nature and Role of Algebra in the K-14 Curriculum: Proceedings of a National Symposium.* National Council of Teachers of Mathematics and Mathematical Sciences Education Board. Center for Science, Engineering, Mathematics, and Engineering Education. Washington, DC: National Academy Press. Available: http://www.nap.edu/catalog.php?record_id=6286.

National Research Council. (1999). *High Stakes: Testing for Tracking, Promotion, and Graduation.* Committee on Appropriate Test Use, J.P. Heubert and R.M. Hauser, Eds. Washington, DC: National Academy Press. Available: http://www.nap.edu/catalog.php?record_id=6336.

National Research Council. (2000a). *Educating Teachers of Science, Mathematics, and Technology: New Practices for the New Millennium.* Committee on Science and Mathematics Teacher Preparation, Center for Education. Washington, DC: National Academy Press. Available: http://www.nap.edu/catalog.php?record_id=9832.

National Research Council. (2000b). *How People Learn: Brain, Mind, Experience, and School: Expanded Edition.* Committee on Development in the Science of Learning, J.D. Bransford, A.L. Brown, and R.R. Cocking, Eds. Commission on Behavioral and Social Sciences and Education. Washington, DC: National Academy Press. Available: http://www.nap.edu/catalog.php?record_id=9853.

National Research Council. (2001a). *Understanding Dropouts: Statistics, Strategies, and High-Stakes Testing.* Committee on Educational Excellence and Testing Equity, A. Beatty, U. Neisser, W.T. Trent, and J.P. Heubert, Eds. Board on Testing and Assessment. Center for Education, Division of Behavioral and Social Sciences and Education. Washington, DC: National Academy Press. Available: http://www.nap.edu/catalog.php?record_id=10166.

National Research Council. (2001b). *Adding It Up: Helping Children Learn Mathematics.* Mathematics Learning Study Committee, J. Kilpatrick, J. Swafford, and B. Findell, Eds. Center for Education, Division of Behavioral and Social Sciences and Education. Washington, DC: National Academy Press. Available: http://www.nap.edu/catalog.php?record_id=9822.

National Research Council. (2003). *Assessment in Support of Instruction and Learning: Bridging the Gap Between Large-Scale and Classroom Assessment—Workshop Report.* Committee on Assessment in Support of Instruction and Learning. Committee on Science Education K-12 and Mathematical Sciences Education Board. Center for Education, Division of Behavioral and Social Sciences and Education. Washington, DC: The National Academies Press. Available: http://books.nap.edu/catalog.php?record_id=10802.

National Research Council. (2007a). *Taking Science to School: Learning and Teaching Science in Grades K-8.* Committee on Science Learning, Kindergarten Through Eighth Grade, R.A. Duschl, H.A. Schweingruber, and A.W. Shouse, Eds. Center for Education, Division of Behavioral and Social Sciences and Education. Washington, DC: The National Academies Press. Available: http://www.nap.edu/catalog.php?record_id=11625.

National Research Council. (2007b). *Ready, Set, Science! Putting Research to Work in K-8 Science Classrooms.* S. Michaels, A.W. Shouse, and H.A. Schweingruber, Eds. Board on Science Education. Center for Education, Division of Behavioral and Social Sciences and Education. Washington, DC: The National Academies Press. Available: http://www.nap.edu/catalog.php?record_id=11882.

National Research Council. (2009). *Learning Science in Informal Environments: People, Places, and Pursuits.* P. Bell, B. Lewenstein, A.W. Shouse, and M.A. Feder, Eds. Board on Science Education. Center for Education, Division of Behavioral and Social Sciences and Education. Washington, DC: The National Academies Press. Available: http://www.nap.edu/catalog.php?record_id=12190.

National Research Council and Institute of Medicine. (2003). *Engaging Schools: Fostering High School Students' Motivation to Learn.* Committee on Increasing High School Students' Engagement and Motivation to Learn. Board on Children, Youth, and Families. Washington, DC: The National Academies Press. Available: http://www.nap.edu/catalog.php?record_id=10421.

National Science Board. (2008). *Science and Engineering Indicators.* Arlington, VA: National Science Foundation. Available: http://www.nsf.gov/statistics/seind08/.

Noyce, P. (2006). Professional development: How do we know if it works? *Education Week,* September 13. Available: http://noycefdn.org/documents/ProfDev_HowDoWeKnowIfItWorks-EdWeek091306.pdf.

Organisation for Economic Co-operation and Development. (2007). *PISA 2006: Science Competencies for Tomorrow's World. Volume 1: Analysis.* Paris: Author. Available: http://www.oei.es/evaluacioneducativa/InformePISA2006-FINALingles.pdf.

Pasley, J.D. (2002). *The Role of Instructional Materials in Professional Development: Lessons Learned from the LSC Community.* Chapel Hill, NC: Horizon Research. Available: http://www.horizon-research.com/LSC/news/pasley2002.php.

Simpson, M.A., and Banilower, E.R. (2004). *Results of the 2003-04 Study of the Impact of the Local Systemic Change Initiative on Student Achievement in Science.* Chapel Hill, NC: Horizon Research. Available: http://www.horizon-research.com/reports/2004/sps0304.php.

Teachers Network. (2007). *Survey Reveals That Only 1% of Teachers Find No Child Left Behind an Effective Way to Assess the Quality of Schools and 69% Report It's Pushing Teachers Out of the Profession.* Press release, April. Available: http://teachersnetwork.org/aboutus/pressreleases/nclb_survey.htm.

Wolk, R. (2009). Why we're still at risk: The legacy of five faulty assumptions. *Education Week,* April 22. Available: http://www.bigpicture.org/2009/04/why-were-still-at-risk-the-legacy-of-five-faulty-assumptions/.

World Economic Forum. (2006). *Switzerland, Finland, and Sweden Take the Lead in the Rankings of the World Economic Forum's Global Competitiveness Index, but United States Drops.* September 26. Available: http://www.weforum.org/en/media/Latest%20Press% 20Releases/GCRpressrelease06.

Yang, J. (2006). Learners and users of English in China. *English Today, 22,* 3-10. Available: http://journals.cambridge.org/action/displayAbstract;jsessionid=3A6A4946D08F15D 57A69987D43A12FC4.tomcat1?fromPage=online&aid=433481.

Yoon, K.S., Duncan, T., Lee, S.W.-Y., Scarloss, B., and Shapley, K.L. (2007). *Reviewing the Evidence on How Teacher Professional Development Affects Student Achievement.* Institute of Education Sciences, National Center for Education Evaluation and Regional Assistance, Regional Educational Laboratory Southwest. Washington, DC: U.S. Department of Education. Available: http://ies.ed.gov/ncee/edlabs/regions/southwest/pdf/REL_ 2007033.pdf.

Appendix A

Convocation Agenda

BUILDING A VILLAGE: LEARNING FROM AND SUSTAINING
SUCCESSFUL PROGRAMS IN ELEMENTARY SCIENCE EDUCATION

**CONVOCATION ON SUSTAINING EFFECTIVE SCIENCE
EDUCATION PROGRAMS IN SCIENCE FOR GRADES K-8
ARNOLD AND MABEL BECKMAN CENTER
APRIL 29-30, 2009**

Wednesday, April 29

8:15-8:30   **Welcoming Remarks and Overview** (of the Convocation's
            goals and expected outcomes):
            Bruce Alberts, **Moderator** (UC San Francisco) National
                Academy Official
            Jacqueline Dorrance, Executive Director, Arnold and
                Mabel Beckman Foundation
            Susan Hackwood, Executive Director, California Council
                on Science and Technology

8:30-8:45   **The Importance of Hands-On, Inquiry-Centered Science
            Education**
            Bruce Alberts

8:45-9:30   **Experiencing Hands-On, Inquiry-Based STEM
            Education, Part I**
            Sue Neuen, Director of Professional Development,
                California Science Center

Banner designed by Judy Harrington, National Academy of Sciences.

Nancy Chung, Fifth Grade Teacher, Tustin Unified School
District

Observe a lesson with fifth grade students with an emphasis on inquiry-based approaches to teaching and learning.
One or two students will sit at each table with convocation
participants.

9:30-10:15   **Experiencing Hands-On, Inquiry-Based STEM Education,
               Part II. Presentations and Voices of School Children**
               Maureen Allen, Orange Country Department of Education

               Students from grades 2-8 in Orange County who have par-
               ticipated in the Beckman@Science Program will be available
               in a mini "poster session" to display the work that they have
               done in science and to answer your questions.

               Refreshments will be available throughout this session.

10:15-10:45   **Teachers' Reflections on the Morning's Activities and
               General Discussion**
               Nancy Chung
               Susan Pritchard, California Science Teachers Association
               Anne-Marie Bergen, California Teacher Advisory Council

10:45-11:45   **How Can the Research Literature Inform Decisions
               About What Can and Should Be Sustained in Science
               Education for Grades K-8?**
               Jeanne Century, Director of Center for Elementary
               Mathematics and Science Education, University of
               Chicago

               Dr. Century will provide participants with an overview of
               the work that has been done to study and understand sus-
               tainable programs in science education. It will emphasize
               the importance of making evidence-based decisions about
               any programs that might emerge in California.

11:45-12:30   **Lunch**

12:30-1:15   **The National Picture and Its Relevance to the Opportunities and Challenges in California for Nurturing and Sustaining Science Education in Grades K-8**
Kathy DiRanna, WestEd

1:15-2:15   **Facilitated Panel Discussion and Participant Engagement: Some Examples of What's Working in K-8 Science Education**
Bruce Alberts, **Facilitator**
Jacqueline Dorrance, Executive Director, Arnold and Mabel Beckman Foundation: Beckman@Science Program
Carlo Parravano, Executive Director, Merck Institute for Science Education: Merck Science Program
Jeffrey Estes, Washington State LASER Program: Washington State LASER (Leadership and Assistance for Science Education Reform)

This session will examine science education programs that evaluations have deemed to be successful. Questions to be explored in this session for each initiative:

- Brief Description of Each Project
- Data Collected
- Strategies for Sustainability
- Challenges Faced

2:15-2:30   **Refreshment Break and Move to Breakout Sessions**

2:30-4:00   **Facilitated Breakout Sessions:**
Facilitators:
Peg Benzie, Beckman@Science, Orange Unified School District
Susie Crandall, ScienceWorks, Fountain Valley School District
Frank Frisch, Science Teaching and Research (STAR) Institute, Chapman University
Margaret Gaston, Center for the Future of Teaching and Learning
Harry Helling, Crystal Cove State Park Cooperative Association (planning committee member)
Pattie Romero, Beckman@Science Coordinator, Arovista Elementary School

In light of what you experienced in the morning hands-on sessions and heard from the panelists and Jeanne Century, what needs to be in place to sustain high-quality science education for grades K-8? How can we sustain these programs? Your challenge is to look at all the stakeholders and come up with support strategies and sustainability solutions.

Participants will be assigned to breakout sessions so that each session has multiple kinds of expertise and perspectives. Each breakout session will delve more deeply into each of the programs described in the morning, with presenters from that session serving as expert resources. The specific goal of each session will be to focus on sustainability issues.

To promote far-ranging discussions based on a spectrum of expertise and perspectives, each participant has been assigned to a breakout session.

Refreshments will be available in a common area beginning at 3:00.

4:00-4:15   **End Breakout Sessions and Return to Plenary Session**

4:15-5:00   **Reports from Breakout Sessions and General Discussion**
Bruce Alberts

- Beckman@Science Group 1
- Beckman@Science Group 2
- Merck Institute for Science Education
- Washington State LASER

5:00-5:30   **Reception for All Participants**

5:30-6:30   **Dinner for All Participants**

6:30-7:30   **Building a Village: It Actually Takes a State**
Arthur Levine, President, Woodrow Wilson Foundation

This presentation discusses why sustained improvements in education need to occur at the state level, focusing on the leverage and opportunity whole state strategies offer.

Thursday, April 30

8:15-9:15   **Moving Toward a Statewide Vision of Nurturing and Sustaining Science Education in Grades K-8: What Will Be Required?**
Dennis Bartels, Exploratorium
Rena Dorph, Lawrence Hall of Science

9:15-9:45   **Connections with Newly Emerging Initiatives in California:**

              California STEM Network Project Supported by the Gates Foundation and the S.D. Bechtel, Jr. Foundation
Susan Hackwood, California Council for Science and Technology
Susan Elrod, Director, Center for Excellence in Science and Mathematics Education, Cal Poly
Brian Kaplan, Silicon Valley Education Foundation

              Qualitative Examination of the Preparation of Elementary School Teachers to Teach Science in California
Eilene Cross, California Council on Science and Technology

              The National Academies' Study on K-12 Engineering Education: Potential Implications for California
Greg Pearson, National Academy of Engineering

9:45-10:15   **Plenary Discussion: Deeper Exploration of Statewide Goals: Multiple Perspectives and Expertise**
Jay Labov, National Academy of Sciences

The goals of this session are for everyone to

1. discuss in greater detail the other projects that are either under way or planned for California, as described in the previous session, and
2. consider from various perspectives the "required elements" for sustaining effective programs in K-8 science education that were articulated by Dennis Bartels and Rena Dorph.

These discussions will then inform the next set of breakout sessions where participants will discuss specific plans for action with others in their sector of the education community.

10:15-10:30   **Refreshment Break and Move to Breakout Session #2**

10:30-12:15   **Breakout Session #2: Deeper Exploration of Statewide Goals: What Can Each Sector of the Education Community Contribute?**

These breakout sessions will explore how the various sectors represented at this convocation can contribute to nurturing and growing successful programs in science education for grades K-8. The goal will be for participants in each breakout to begin developing models for sustainability of K-8 science education both regionally and across the state.

The following breakout groups will meet:

- Business and Industry
- Foundations
- Education Administration (local and state levels)
- Education Researchers
- Informal Education
- K-12 and Higher Education Faculty

Participants who are not in one of these groups should join the breakout session of their choice.

Refreshments will be available throughout these breakout sessions.

12:15-12:30   **Break—Obtain lunch in dining room and return to the atrium for lunch and final plenary.**

12:30-1:15   **Brief Reports from Each of the Breakout Sessions**

1:15-2:00   **Summary and Closing Remarks—Challenge to Continue the Process**

Additional details about possible next steps for follow-up activities and who might be involved.

Summary by Bruce Alberts, Members of the Organizing Committee, and Convocation Participants

Appendix B

Convocation Participants

Bruce Alberts
Professor
University of California, San
    Francisco

Maureen Allen
Consultant/Instructor
Beckman@Science
Los Alamitos, CA

Denise Antrim
K-12 Science Coordinator
Orange County Department of
    Education

Peter Arvedson
High School Science Teacher
California Teacher Advisory
    Council
La Puente, CA

Michael Barker
Administrator
Tustin Unified School District
Tustin, CA

Lynn Baroff
Executive Director
California Space Education &
    Workforce Institute
Pasadena, CA

Dennis Bartels
Executive Director
Exploratorium
San Francisco, CA

Raymond Bartlett
Senior STEM Consultant
Teaching Institute for Essential
    Science
Columbia, MD

Patricia Beckman
Board of Directors
Arnold and Mabel Beckman
    Foundation
Newport Beach, CA

Margaret Benzie
Science Coordinator, Orange
　Unified
Beckman@Science
Orange, CA

Anne Marie Bergen
Chair
California Teacher Advisory
　Council
California Council on Science and
　Technology
Oakdale, CA

Judy Bishop
Educational Director
EcoCenter-Renewable Energy
　Education
San Diego, CA

Joan Bissell
Director, Teacher Education
California State University
Long Beach, CA

George Bo-Linn
Chief Program Officer
Gordon and Betty Moore
　Foundation
Palo Alto, CA

Susan Brady
Department Head
Center for Science and
　Engineering Education
Lawrence Berkeley National
　Laboratory
Berkeley, CA

Bonnie Brunkhorst
Professor
California State University
San Bernardino, CA

Herb Brunkhorst
Professor/Department Chair
California State University
San Bernardino, CA

Winnie Callahan
Director for Business Education
　Government Health Initiative
Information Sciences Institute-USC
Marina Del Rey, CA

Richard Cardullo
Professor and Chair of Biology
University of California
Riverside, CA

Janet Carlson
Executive Director
BSCS (Biological Sciences
　Curriculum Study)
Colorado Springs, CO

Betty Carvellas
Teacher Leader, Teacher Advisory
　Council
National Academies
Colchester, VT

Robin Casselman
Project Director, FOCUS MSP
University of California
Irvine, CA

Jeanne Century
Director of Science Education
Center for Elementary Mathematics
　and Science Education
Chicago, IL

Caleb Cheung
Science Program Manager/CCTC
    Chair
Oakland Unified School District
California Commission for Teacher
    Credentialing
Oakland, CA

Nancy Chung
Fifth Grade Teacher
Tustin Unified School District
Irvine, CA

Linda Clinard
Literacy Specialist
Center for Education Partnerships
University of California
Irvine, CA

Angelo Collins
Executive Director
Knowles Science Teaching
    Foundation
Moorestown, NJ

John Collins
Scientist
Maxwell Sensors
Irvine, CA

Andrew Coulson
President, Education Division
MIND Research Institute
Santa Ana, CA

Susie Crandall
Director
ScienceWorks
Fountain Valley, CA

Linda Crans
University Trustee/Business
    Consultant
Western University of Health
    Sciences
Diamond Bar, CA

Eilene Cross
Education Consultant
California Council on Science and
    Technology
Pleasanton, CA

Dean Davis
Senior Principal Scientist/Engineer
Boeing Phantom Works
Manhattan Beach, CA

Kathy DiRanna
Director
K-12 Alliance/WestEd
Santa Ana, CA

Rena Dorph
Research and Evaluation Director
Lawrence Hall of Science
University of California
Berkeley, CA

Jacqueline Dorrance
Executive Director
Arnold and Mabel Beckman
    Foundation
Irvine, CA

Marilyn Edling
Project Manager
California STEM Innovation
    Network
Granite Bay, CA

Janet English
Teacher on leave
Director, Educational Services
KOCE-TV, PBS
Lake Forest, CA

Jeff Estes
Manager
Science & Engineering Education
Washington State LASER
Richland, WA

Dorothy Fleisher
Program Director
W.M. Keck Foundation
Los Angeles, CA

Darryl Flick
Director of Development
BayBio Institute
South San Francisco, CA

Sharon Freeburn
Director
Science and Technology Teacher
    Resource Center
San Diego, CA

Frank Frisch
Professor
Department of Biological Sciences
Orange, CA

Matt Gardner
Chief Executive Officer
BayBio
South San Francisco, CA

Margaret Gaston
President
Center for the Future of Teaching
    and Learning
Santa Cruz, CA

Jim Gentile
President and Chief Executive
    Officer
Research Corporation
Tucson, AZ

Dean Gilbert
Science Consultant
Los Angeles County Office of
    Education
Downey, CA

Susan Hackwood
Executive Director
California Council on Science and
    Technology
Riverside, CA

Susie Hakansson
Executive Director
California Mathematics Project
Los Angeles, CA

Virgil Hammon
Pre-College Education
NASA/Jet Propulsion Laboratory
Pasadena, CA

James Hamos
Program Director
National Science Foundation
Arlington, VA

William Harris
President and Chief Executive
    Officer
Science Foundation Arizona
Phoenix, AZ

Susan Harvey
Senior Program Officer
S.D. Bechtel, Jr. Foundation
San Francisco, CA

Harry Helling
President and Chief Executive
    Officer
Crystal Cove Alliance
Newport Coast, CA

Laura Henriques
Professor, Science Education
California State University
Long Beach, CA

Don Hicks
Executive Director
TechAmerica
Irvine, CA

Cynthia Jolly
STEM Consultant
TIES (Technology and Information
    Educational Services)
Raleigh, NC

Juliana Jones
Teacher, California Teacher
    Advisory Council
National Teacher Advisory Council
Longfellow Middle School
Berkeley, CA

Brian Kaplan
Director
Silicon Valley Education
    Foundation
San Jose, CA

Elaine Keeley
Administrator, Curriculum and
    Instruction
Orange County Department of
    Education
Santa Ana, CA

John Kenney
Education Partnerships Manager
PASCO Scientific
Roseville, CA

Gay Krause
Director
Krause Center for Innovation at
    Foothill College
Los Altos Hills, CA

Jay Labov
Senior Advisor for Education and
    Communication
National Academy of Sciences
Washington, DC

Phil LaFontaine
Director
California Department of
    Education
Sacramento, CA

Arthur Levine
Woodrow Wilson Foundation
Princeton, NJ

Ethan Lipton
Professor
California State University
Los Angeles, CA

Maria Alicia Lopez Freeman
Executive Director
California Science Project
Los Angeles, CA

Lindsey Malcom
Assistant Professor, Higher
    Education
Graduate School of Education
University of California
Riverside, CA

Debra Mauzy-Melitz
Director, STEM Faculty Outreach
University of California
Irvine, CA

Lisa McClellan
Mentor Teacher
Cotsen Family Foundation/
Los Angeles Unified School
    District
Midway City, CA

Alan McCormack
Professor of Science Education
San Diego State University
San Diego, CA

John McDonald
Consultant
Stone's Throw Strategic
    Communications
Manhattan Beach, CA

Judy Miner
President
Foothill College
Los Altos Hills, CA

Rich Montgomery
Principal
Tustin Unified School District
Irvine, CA

Janice Morrison
Executive Director
Teaching Institute for Excellence
    in STEM
Baltimore, MD

Peter Murray
Dean
Foothill College
Los Altos Hills, CA

Suzanne Nakashima
Teacher
California Teacher Advisory
    Council
Yuba City, CA

Marwan Nasralla
President and Chief Executive
    Officer
IVD Technologies
Santa Ana, CA

Sue Neuen
Director of Professional
    Development
California Science Center
Los Angeles, CA

Steve Olson
Science Writer
Bethesda, MD

Carlo Parravano
Executive Director
Merck Institute for Science
    Education
Rahway, NJ

Greg Pearson
Senior Program Officer
National Academy of Engineering
Washington, DC

James Pellegrino
Professor
University of Illinois at Chicago
Chicago, IL

Angela Phillips Diaz
Special Assistant to the Chancellor
University of California
Riverside, CA

Donna Pozzi
Chief of Interpretation and
    Education
California State Parks
Sacramento, CA

Susan Pritchard
Teacher
California Science Teachers
    Association
California Teacher Advisory
    Council
La Habra, CA

Stephanie Ramirez
Program Associate
Gordon and Betty Moore
    Foundation
Palo Alto, CA

Denise Reid
Teacher
Signal Hill Elementary
Signal Hill, CA

Ron Rohovit
Deputy Director of Education
California Science Center
Los Angeles, CA

Pattie Romero
Teacher
Brea Olinda School District
Brea, CA

Kathleen Roth
Incoming Director
Center for Professional
    Development
BSCS (Biological Sciences
    Curriculum Study)
Colorado Springs, CO

Steve Schneider
Senior Program Director
Mathematics and Science
WestEd
Redwood City, CA

Kandarp Shah
Graduate Student
University of California
Irvine, CA

Therese Shanahan
Codirector
California Science Project at Irvine
Irvine, CA

Barbara Shannon
Educator/Principal
California Teacher Advisory
    Council
Monrovia, CA

Patrick Shields
Director, Center for Education
    Policy
SRI International
Menlo Park, CA

Philip Smith
Director
Space Grant Education and
    Enterprise Institute
San Diego, CA

Gerald Solomon
Executive Director
Samueli Foundation
Corona del Mar, CA

Roslyn Soto
Director, STEM Major
    Development
University of California
Irvine, CA

Mark St. John
Evaluator
Inverness Research
Inverness, CA

Elizabeth Stage
Director
Lawrence Hall of Science
University of California
Berkeley, CA

Art Sussman
Senior Project Director
WestEd
Richmond, CA

Nancy Taylor
K-12 Science Cordinator
San Diego County Office of
    Education
San Diego, CA

Jerry Valadez
Director
Central Valley Science Project
Fresno, CA

Deena Vela
Teacher on Special Assignment
Tustin Unified School District
Tustin, CA

Soo Venkatesan
Program Officer
S.D. Bechtel, Jr. Foundation
San Francisco, CA

Robert Willis
Teacher
District of Columbia Public Schools
Fort Washington, MD

Janet Yamaguchi
Vice President, Education
Discovery Science Center
Santa Ana, CA

Russ Yarrow
General Manager
California Corporate Affairs
Chevron
San Ramon, CA

Appendix C

Biographical Sketches of
Presenters and Facilitators

Bruce Alberts, a biochemist with a strong commitment to the improvement of science education, began service as editor-in-chief of *Science* in March 2008. Alberts was also a long-time professor in the Department of Biochemistry and Biophysics at the University of California, San Francisco, and served two 6-year terms as the president of the National Academy of Sciences (NAS). Alberts is one of the original authors of *The Molecular Biology of the Cell*, a textbook now in its fifth edition. For the period 2000 to 2009, he served as cochair of the InterAcademy Council, an organization in Amsterdam governed by the presidents of 15 national academies of sciences and established to provide scientific advice to the world. Widely recognized for his work in the fields of biochemistry and molecular biology, Alberts has earned many honors and awards, including 15 honorary degrees. He currently serves on the advisory boards of more than 25 nonprofit institutions, including the Gordon and Betty Moore Foundation and the Lawrence Berkeley National Laboratory.

Maureen Allen is recently retired from the Department of Education in Orange County, California, as a program specialist for K-12 science education. During her 8 years at the county office, she worked with the 28 school districts in Orange County conducting science staff development, writing science curriculum, and helping to coordinate several different grants, as a partner with school districts, universities, businesses, and professional organizations. She has also worked with the Beckman@Science Program as a teacher leader, kit trainer, and coordinator for Irvine Unified School

District and represented the Orange County Department of Education. Previously she was a K-6 science resource specialist and middle school teacher for Irvine Unified School District for 22 years. She also developed its curriculum and coordinated its after-school science program and helped to coordinate the Astounding Inventor's Competition. She was twice selected as Irvine's Teacher of the Year and was a semifinalist for the California State Science Teacher of the Year. Statewide, Allen worked on two projects of the Curriculum and Instruction Steering Committee: the Strategic Science Teaching Workbook and a statewide science staff development program, "It's All About the Electron." She was the consulting author for the Scott Foresman, Science Series, K-6, Discover the Wonder and has coauthored five AIMS Education Foundation books.

Dennis Bartels is executive director of the Exploratorium—San Francisco's museum of science, art, and human perception. He studies curriculum reform, teacher professional development, technology in education, learning theory, and organizational change. Previously he was president of TERC, an education research and development center based in Cambridge, MA. Prior to 2001, Bartels directed the Center for Teaching and Learning at the Exploratorium. He was project director of the South Carolina Statewide Systemic Initiative and directed the development of the state curriculum frameworks there. He has served on several committees, advisory boards, and review panels for the National Science Foundation and other education organizations, including the Merck Institute for Science Education and the Cisco Learning Institute. He is a fellow of the American Association for the Advancement of Science and the American Educational Research Association. He has a Ph.D. in education administration and policy analysis from Stanford University and completed his undergraduate degree at the University of North Carolina, Chapel Hill.

Anne Marie Bergen is the coordinator of and teacher in the Oakdale Joint Unified District Science Program. Over the past 8 years she has created many programs—Passport to Science, From the River to the Tap, Salmon project—to link students and teachers to the world around them. With an emphasis on hands-on science experiences, she provides demonstration labs and professional development to over 2,500 students and 100 teachers. She coordinates with area scientists to bring their expertise to elementary through high school students. Her honors include Stanislaus County Teacher of the Year 2002, California Teacher of the Year 2003, and the AMGEN Award for Science Teaching Excellence. Most recently she was honored as the 2008 Cal Poly Alumni of the Year for the College of Science and Mathematics. She is chair of the California Teacher Advisory

Council (Cal TAC), working to improve science, mathematics, engineering, and technology education in California.

Jeanne Century is director of science education and research and evaluation at the University of Chicago's Center for Elementary Mathematics and Science Education (CEMSE). She directs several research and evaluation efforts and leads CEMSE's other science education and research work. Previously, she was a senior project director in the Center for Science Education at the Education Development Center in Newton, Massachusetts. Century has spent the majority of her 20-year career working in and with urban schools and large urban school districts around the country. She has developed comprehensive science instructional materials for the elementary and middle school levels as well as materials for informal settings. She has also conducted a range of professional development for teachers and school and district administrators around the country and provided technical assistance and strategic planning support to the leadership of science education improvement efforts at the school, district, and state levels. Her research and evaluation efforts have focused on the impact of inquiry-based science instruction, strategies for improving the use of research and evaluation, the sustainability of reform efforts, and measurement of innovation fidelity and use. Most recently, Century served on the education policy transition team and Department of Education agency review transition team for the Obama-Biden administration, focusing on STEM education and education research and development. She has a K-8 teaching certificate and an undergraduate degree in general science from Brandeis University and master's and doctorate degrees in science education curriculum and teaching from Boston University.

Nancy Chung is a fifth grade teacher in her eleventh year of teaching. She considers herself very fortunate to be teaching at the "world class" Hicks Canyon Elementary School in Irvine, California, where she is able to live out her dream and passion for teaching on a daily basis. With a bachelor of fine arts degree in art education from California State University, Fullerton, and a multiple-subject credential and Masters of Education from Biola University, she has served as a master teacher, the district math coach, grade level lead, and Hicks Canyon Quad Mathalon coach. She has been an active member of the Tustin Unified School District Science Steering Committee, the Professional Learning Community Committee, the Response to Instruction Committee, and the California Distinguished School Committee. Committed to hands-on learning, she encourages her students to become lifelong learners.

Susan Crandall is coordinator of the ScienceWorks Consortium, whose mission is to combine the resources of the Fountain Valley, Huntington Beach City, and the Westminster School Districts, as well as all other communities, to support all students in attaining high levels of scientific literacy. She develops and delivers staff development in science to K-8 teachers in 36 schools and supervises the Materials Resource Center, which refurbishes and circulates over 4,500 science modules to 55 schools in 5 school districts. Budget constraints have made establishing partnerships with community and corporate leaders to sustain the programs of the ScienceWorks Consortium a major focus for the past three years. As a former educational consultant for Scholastic, Houghton Mifflin, Riverside, and Laidlaw Brothers, Crandall has created and presented staff development workshops in all K-12 curriculum areas and assisted school districts with implementation of state standards and frameworks. With over 30 years in the classroom, she has completed extensive work integrating reading, language arts, and mathematics with science. Crandall has K-8 and K-12 teaching certificates from several states and administrative certificates in California. She is a graduate of the University of New Hampshire and has a master's in educational administration from United States International University, San Diego.

Eilene Cross is an education consultant with the California Council on Science and Technology. From 1994 to 2007 she worked for Sandia National Laboratory/California as the program manager and head mentor of the intern program. She developed all aspects of the program, including recruitment, selection, project supervision, professional development, and evaluation. She established intern institutes in the fields of combustion research, biotechnology, nanotechnology, and advanced material development. She also served as a mentor in Sandia's outreach program that honors and supports high school girls who are high achievers and rising stars in science and mathematics. She is a lead judge for the Intel International Science and Engineering Fair in the California tri-valley region. While at Sandia, Cross was also a research material scientist, working to develop room temperature radiation detectors. For this research, she received the Discover Magazine Innovation Award in 1997 and the R&D 100 Award in 1998, 1999, and 2001. From 1976 to 1994 she worked for EG&G in Santa Barbara as a senior scientist. Eilene has a B.S. in biology from Monmouth College and continued her studies in the M.A./Ph.D. program in geography at the University of California, Santa Barbara.

Kathy DiRanna is the statewide director of the K-12 Alliance, which focuses on school and department-wide change by providing programs that address content, instructional strategies, assessment, and leadership. She served as a principal investigator or project director for the California Systemic Initiative, the Center for the Assessment and Evaluation of Student Learning (CAESL), and Science Partnerships for Articulation and Networking (SPAN). She has helped shape California's science reform efforts for the past 23 years and continues to be an advocate for the reform efforts through the California Mathematics and Science Partnership Program. She has served as the mentor coordinator for the National Academy of Science and Mathematics Education, codeveloped the professional development design of the BSCS (Biological Sciences Curriculum Study) Science Curriculum Implementation Center, the Full Option Science System Leadership Academy, and the Using Data Project. She serves on a variety of advisory boards, has been a consultant on instructional materials and multimedia productions, is a featured speaker at state and national conferences, and served as the program coordinator for National Science Teachers Association's 2006 national conference. She is the coauthor of several publications and has received awards that include the Cal Alive Educator of the Year; WestEd's Paul Hood Individual Award and Paul Hood Team Award for making significant contributions to the field; and the California Science Teachers Association's highest honor, the Margaret Nicholson Award for distinguished service to science education.

Rena Dorph is the director of the Center for Research, Evaluation, and Assessment (REA) at the Lawrence Hall of Science (LHS) at the University of California, Berkeley. REA contributes to excellence, equity, and innovation in science and mathematics education by conducting applied research, evaluating the quality and impact of educational materials and programs, and developing assessments that measure knowledge and learning in science and mathematics. In this capacity, Dorph provides leadership and support for the LHS community as well as for organizations that contract with the REA Center. She has worked in the field of education research and evaluation for 15 years. Previously she worked as director for Research, Policy, and Technology in the Teacher Education and Professional Development unit of the University of California Office of the President; as an education consultant for California schools, districts, and county offices of education; as the lead researcher and coordinator for the School Restructuring Study based at the University of California, Berkeley; and for the National Center for Restructuring Education, Schools, and Teaching at Columbia University/Teachers College. She has a Ph.D. in education policy, organization, measurement, and evaluation from the University of California, Berkeley, an M.A. in the sociology of education from Columbia University/Teachers College, and a B.A. in psychology from the University of California, Davis.

Jacqueline Dorrance is executive director of the Arnold and Mabel Beckman Foundation, one of the largest scientific foundations in the country. She is responsible for its overall management, including development of the current and long-range objectives, investment management, staff management and development, program development and management, public relations, human resource management, budget development and management, regulatory and financial oversight, and special event generation. From 1992 to 1994 she worked as chief administrative officer of ARCO, Los Angeles. Her responsibilities included managing and directing the chairman's staff, public relations, state and local government liaison activities, and board relations. Previously she worked for ARCO in Denver. She has also worked in the textile industry in the United States and China, working to improve working conditions in local Chinese factories, quality control, pricing, office space planning, and development. She attended the University of Colorado, Boulder, majoring in business management.

Jeffrey Estes is manager of the Office of Science Education and Community Relations at the Pacific Northwest National Laboratory (PNNL). In this role, he oversees a laboratory-based science and engineering education program that spans elementary through undergraduate school. The program focuses on linking the laboratory's human, financial, and technical resources with elementary, secondary, and postsecondary schools to help improve the teaching and learning of science, technology, engineering, and mathematics (STEM). He oversees a portfolio of (1) workforce development programs for high school and undergraduate students, (2) laboratory-based professional development programs for precollege teachers and college/university faculty, (3) science education reform projects in K-12 schools, and (4) outreach programs involving local schools and community organizations focused on STEM education. He is the coleader of the Washington State Leadership and Assistance for Science Education Reform (LASER) Regional Partnership, providing the technical assistance required by school districts to initiate, implement, and sustain reform efforts. He manages a set of professional development, outreach, and technical assistance activities, provided by PNNL, that support LASER's efforts. Now in his 34th year in education, Estes had worked in formal and informal education settings including elementary school, middle school, museum education, and a U.S. Department of Energy national laboratory.

Frank Frisch is a physiologist in the Department of Biological Sciences at Chapman University. His research and teaching areas are in the areas of metabolism. His current work is in the elucidation of bone metabolism, specifically ways in which rates of osteoporosis and osteopenia can be attenuated. He teaches courses in comparative and human physiology and anatomy. He has a long-time interest in science education and in addition to his research grants is the coprincipal investigator for a grant from the California Postsecondary Education Commission focusing on science education for K-2 teachers, a California Mathematics and Science Partnership Collaboration for Support in Science Partnership for fourth and fifth grade teachers, and a lead scientist for another grant serving sixth and seventh grade teachers of science. His view is that the entire society must improve in science literacy in order to promote and advance civilization.

Margaret Gaston is president and executive director of the Center for the Future of Teaching and Learning, a public policy organization focused on strengthening California's teacher workforce. Prior to establishing the center in 1994, she was a special consultant to philanthropic organizations, advising foundation program officers, presidents, and board members on grant making to public education and education support entities. She has worked as a consultant and administrator for the California Department of Education, overseeing the School Improvement Program, the School Program Quality Review, Community Education, and other school reform efforts. At the local level, she has been an elementary school teacher, a categorical programs coordinator, and a high school vice principal. As San Diego State University's distinguished alumna in the field of education in 2007, she received its Monty Award. She was appointed to the California Commission on Teacher Credentialing and now serves as vice chair.

Susan Hackwood is executive director of the California Council on Science and Technology (CCST) and professor of electrical engineering at the University of California, Riverside. CCST is a not-for-profit corporation comprised of 150 top science and technology leaders sponsored by the key academic and federal research institutions in California, which advises the state on all aspects of science and technology. In 1990, she became the founding dean of the Bourns College of Engineering at the University of California, Riverside. Her current research interests include science and technology policy, innovation mechanisms, distributed asynchronous signal processing, and cellular robot systems. She is a fellow of the IEEE and the American Association for the Advancement of Science and holds honorary degrees from Worcester Polytechnic Institute and DeMontfort University, UK. She has worked extensively with industry, academic, and government partnerships to identify policy issues of importance. She is currently involved with science and technology development in California, the United States, Mexico, Ireland, Taiwan, Vietnam, and Costa Rica. She has a Ph.D. in solid state ionics from DeMontfort University. Before joining academia, she was department head of device robotics technology research at AT&T Bell Labs. In 1984 she joined the University of California, Santa Barbara, as professor of electrical and computer engineering and was founder and director of the National Science Foundation's Engineering Research Center for Robotic Systems in Microelectronics.

Harry Helling is president and chief executive officer of the Crystal Cove Alliance (CCA) on the Newport Coast in California. CCA is a nonprofit partner of the California State Parks with the mission of developing innovative educational programs, coordinating ecological research that informs resource management decisions, and raising funds for restoration and conservation in the Parks' historic district. He recently left the Ocean Institute, Dana Point, California, after 24 years as executive vice president in charge of education and research. He founded the Center for Cooperation for Research and Education in 2002 and has developed effective education and outreach partnerships with over 20 university research groups. He is principal investigator of SeaTech: Underserved Teens Hooked on Ocean Technology, which trains teens at Boys & Girls Clubs to contribute to whale bioacoustics research at the Scripps Institution of Oceanography.

Jay Labov is senior advisor for education and communication for the National Academy of Sciences (NAS) and the National Research Council (NRC). He has been the study director and responsible staff officer for 10 NRC reports that have focused on many aspects of STEM education, including teacher education, advanced study for high school students, and evaluating and improving undergraduate education. He was the coprincipal investigator for a multiyear grant from the National Science Foundation to offer workshops to grantees of its Math/Science Partnership Initiative to enable them to better understand and implement the recommendations in NRC reports and other research literature on their own work. He also directed a committee of the NAS and the Institute of Medicine that authored *Science, Evolution, and Creationism* (2008) and continues to oversee efforts at the NAS to confront challenges to the teaching of evolution in the nation's public schools. Previously he was a member of the faculty in the Department of Biology at Colby College. He has a B.S. in biology from the University of Miami and an M.S. in zoology and a Ph.D. in biological sciences from the University of Rhode Island. He was awarded a Kellogg National Fellowship (1988-1990), elected as a fellow in education of the American Association for the Advancement of Science in 2005, and appointed as a Woodrow Wilson visiting fellow in 2008.

Arthur Levine is the sixth president of the Woodrow Wilson National Fellowship Foundation in Princeton, New Jersey. Previously, he was president and professor of education at Teachers College, Columbia University. He also served as chair of the higher education program, chair of the Institute for Educational Management, and senior lecturer at the Harvard Graduate School of Education. He was president of Bradford College (1982-1989) and senior fellow at the Carnegie Foundation and the Carnegie Council for Policy Studies in Higher Education (1975-1982). His most recent book is *When Hope and Fear Collide: A Portrait of Today's College Student* (with Jeanette S. Cureton). Much of his research and writing in recent years has focused on increased educational opportunity and strengthening schools of education. He has received numerous honors, including a Guggenheim fellowship and a Carnegie fellowship, as well as the American Council on Education's Book of the Year award (for *Reform of Undergraduate Education*), the Educational Press Association's annual award for writing (three times), and 19 honorary degrees. He has a bachelor's degree from Brandeis University and a Ph.D. from the State University of New York, Buffalo.

Sue Neuen is director of professional development in Orange County at the California Science Center. In that capacity, she leads the Science@OC initiatives, which are designed to provide an infrastructure to deliver comprehensive, research-based sustainable programs to reform how students learn, teachers teach, and the community supports science education. Support for teachers of middle school science—"the gatekeepers to success in high school science"—is the current focus of the Orange County Middle School Initiative through Science@OC. She has served as a resource team leader for 26 of the National Science Resources Center's National and Regional Leadership Assistance for Science Education Reform Strategic Planning Institutes from 1997 to 2009. She is directing the formation of the California STEM (science, technology, engineering, and mathematics) Coalition, an affiliate program of the National Alliance for State Science and Math Coalitions, consisting of education, business, and public policy leaders. As associate vice president for the Americas for MANCEF, the Micro and Nanotechnology Commercialization Education Foundation, she is an international speaker on global competitiveness, STEM education, and the 21st century workforce. She is cofounder of Neusoli, an education architectural firm designed to educate the public and build the nanotechnology workforce pipeline. From 1996 to 2003 she was director of the Einstein Project in Green Bay, Wisconsin; under her leadership, the project grew from a county program to a national model, nonprofit organization providing services to 69 school districts statewide.

Carlo Parravano is executive director of the Merck Institute for Science Education, where he is responsible for the planning, development, and implementation of numerous initiatives to improve science education. Previously he was professor of chemistry and chair of the Division of Natural Sciences at the State University of New York (SUNY) at Purchase and director of the Center for Mathematics and Science Education of the SUNY/Purchase-Westchester School Partnership. He is a fellow of the American Academy for the Advancement of Science and a national associate of the National Academies. He is an active member in a number of professional organizations, and he is the recipient of the SUNY Chancellor's Award for Excellence in Teaching; the National Science Teachers Association's Distinguished Service to Science Education Award; the Keystone Center's Leadership in Education Award; Rutgers University's Distinguished Leader Award; and the Education 500 Leadership Award of the Institute for Education Excellence and Entrepreneurship. His research experience has been in the area of molecular beam studies of excited atoms and molecules and the application of physical chemical techniques to the solution of biochemical and environmental problems. He has a B.A. in chemistry from Oberlin College and a Ph.D. in physical chemistry from the University of California, Santa Cruz.

Greg Pearson is a senior program officer in the National Academy of Engineering (NAE). In that role, he develops and manages new areas of activity related to K-12 engineering education, technological literacy, and the public understanding of engineering. He is currently at work on two projects: "Understanding and Improving K-12 Engineering in the United States" and "Exploring Content Standards for Engineering Education in K-12." He was the study director for a project that resulted in the 2008 publication *Changing the Conversation: Messages for Improving Public Understanding of Engineering* and was coeditor of the reports *Tech Tally: Approaches to Assessing Technological Literacy* (2006) and *Technically Speaking: Why All Americans Need to Know More About Technology* (2002). In the late 1990s, Pearson oversaw NAE and National Research Council reviews of technology education content standards developed by the International Technology Education Association. He works with colleagues inside and outside the National Academies on a variety of projects involving K-12 science, mathematics, technology, and engineering education and the public understanding of engineering and science. He has an undergraduate degree in biology from Swarthmore College and a graduate degree in journalism from American University.

Susan M. Pritchard is a middle school science teacher with the La Habra City School District in California. A passionate science advocate, she encourages students to succeed both in academics and in life choices and does not allow them to accept failure. She will keep students after school (with parental support) until the work expected is completed to the best of their abilities. From achieving her national board certification in early adolescence science to continuing her pursuit of knowledge through college science courses, Pritchard is always seeking ways to better explain science concepts to her students through exciting inquiry-based investigations. Borrowing ideas from courses and workshops and adapting them to her eighth grade curriculum allows her to enhance student-learned outcomes in a motivating and positive classroom atmosphere. With a belief that a science-literate citizenry will greatly benefit the world's future, her goal is to have her students become science ambassadors who can reach out to their community, teach people about science concepts, and further their own lifelong learning in the process. As a member of the California Teachers Advisory Council (Cal TAC) and the California Science Teachers Association (CSTA) Board of Directors for over 7 years, currently as past president, she promotes science education throughout the state. She has a Ph.D. from Claremont Graduate School, an M.S. from California State University, Fullerton, and a B.A. from the University of California, Irvine.

Pattie Romero is a fifth grade teacher in the Brea-Olinda Unified School District. With 21 years of experience, she is a two-time Teacher of the Year recipient. She has worked with the Beckman@Science Program since its inception in all capacities of the program, including teacher leader, science education fellow, and program director. Her responsibilities have included running a county-wide materials center and a professional development program. She has an M.A. in teaching science from California State University, Fullerton, and an administrative credential from the University of California, Irvine.

Appendix D

Summary of Selected
National Academies Reports

Some key reports from the National Academies are summarized in this appendix (listed in reverse chronological order). The Executive Summary for each report is available for downloading without cost at the URL listed.

Learning Science in Informal Environments: People, Places, and Pursuits (2009)
http://books.nap.edu/catalog.php?record_id=12190.

Taking Science to School: Learning and Teaching Science in Grades K-8 (2007)
http://www.nap.edu/catalog.php?record_id=11625.

Ready, Set, Science! Putting Research to Work in K-8 Science Classrooms (2007)
http://www.nap.edu/catalog.php?record_id=11882

Enhancing Professional Development for Teachers: Potential Uses of Information Technology (2007)
http://www.nap.edu/catalog.php?record_id=11995

Tech Tally: Approaches to Assessing Technological Literacy (2006)
http://www.nap.edu/catalog.php?record_id=11691

Technically Speaking: Why All Americans Need to Know More About Technology (2002)
http://www.nap.edu/catalog.php?record_id=10250

Adding It Up: Helping Children Learn Mathematics (2001)
http://www.nap.edu/catalog.php?record_id=9822

Inquiry and the National Science Education Standards: A Guide for Teaching and Learning (2000)
http://www.nap.edu/catalog.php?record_id=9596

How People Learn: Brain, Mind, Experience, and School—Expanded Edition (2000)
http://www.nap.edu/catalog.php?record_id=9853

Educating Teachers of Science, Mathematics, and Technology: New Practices for the New Millennium (2000)
http://www.nap.edu/catalog.php?record_id=9832

Learning Science in Informal Environments: People, Places, and Pursuits (2009)

General Description

Science is shaping people's lives in fundamental ways. Efforts to enhance scientific capacity typically target schools and focus on such strategies as improving science curriculum and teacher training and strengthening the science pipeline. What is often overlooked or underestimated is the potential for science learning in nonschool settings, where people actually spend the majority of their time.

This report examines the potential of nonschool settings for science learning. The authoring committee assessed the evidence of science learning across settings, learner age groups, and over varied spans of time; they identified the qualities of learning experiences that are special to informal environments and those that are shared (e.g., with schools); and proposed an agenda for research and development The committee examined the places where science learning occurs as well as crosscutting features of informal learning environments. The "places" include everyday experiences—like hunting, walking in the park, watching a sunrise—designed settings—such as visiting a science center, zoo, aquarium, botanical garden, planetarium—and programs—such as afterschool science or environmental monitoring through a local organization. Cross-cutting features that shape informal environments include the role of media as a context and tool for learning and the opportunities these environments provide for inclusion of culturally, socially, and linguistically diverse communities.

The report is focused on the following issues:

- **Defining Appropriate Outcomes:** To understand whether, how, or when learning occurs, good outcome measures are necessary, yet efforts to define outcomes for science learning in informal settings have often been controversial. At times, researchers and practitioners have adopted the same tools and measures of achievement used in school settings. Yet traditional academic achievement outcomes are limited. Although they may facilitate coordination between informal environments and schools, they fail to reflect the defining characteristics of informal environments. The challenge of developing clear and reasonable goals for learning science in informal environments is compounded by the real or perceived encroachment of a school agenda on such settings. This has led some to eschew formalized outcomes altogether and to embrace learner-defined outcomes instead. The authoring committee's view is that it is unproductive to blindly adopt either purely academic goals or purely subjective learning goals. Instead, the committee prefers a third course that combines a variety of specialized science learning goals used in research and practice.

- **Strands of Science Learning:** The committee proposed a "strands of science learning" framework that articulates science-specific capabilities supported by informal environments. It builds on the framework developed for K-8 science learning in *Taking Science to School* (see below) and in the growing body of evidence that learning also occurs in these environments across the strands. They added two additional strands—Strands 1 and 6—which are of special value in informal learning environments.

Learners in informal environments:

Strand 1: Experience excitement, interest, and motivation to learn about phenomena in the natural and physical world.

Strand 2: Come to generate, understand, remember, and use concepts, explanations, arguments, models, and facts related to science.

Strand 3: Manipulate, test, explore, predict, question, observe, and make sense of the natural and physical world.

Strand 4: Reflect on science as a way of knowing; on processes, concepts, and institutions of science; and on their own process of learning about phenomena.

Strand 5: Participate in scientific activities and learning practices with others, using scientific language and tools.

Strand 6: Think about themselves as science learners and develop an identity as someone who knows about, uses, and sometimes contributes to science.

The strands are distinct from, but overlap with, the science-specific knowledge, skills, attitudes, and dispositions that are ideally developed in schools.

- **Broadening Participation:** There is a clear and strong commitment among researchers and practitioners to broadening participation in science learning. Efforts to improve inclusion of individuals from diverse groups are under way at all levels and include educators and designers, as well as learners themselves. However, it is also clear that laudable efforts for inclusion often fall short. Research has turned up several valuable insights into how to organize and compel broad, inclusive participation in science learning. The report provides an array of conclusions about ways to broaden inclusion.

Relevance to Convocation Participants

This report is important to convocation participants because it stresses the broad possibilities for people to learn about science both inside and outside school settings. Informal education, both in its own right and when well integrated with formal education, has the potential to reach and engage many more students, and especially younger children, than either setting alone.

Recommendations

Exhibit and Program Designers

Exhibit and program designers play an important role in determining what aspects of science is reflected in learning experiences, how learners engage with science and with one another, and the type and quality of educational materials that learners use.

Recommendation 1: Exhibit and program designers should create informal environments for science learning according to the following principles. Informal environments should

- Be designed with specific learning goals in mind (e.g., the strands of science learning).
- Be interactive.
- Provide multiple ways for learners to engage with concepts, practices, and phenomena within a particular setting.
- Facilitate science learning across multiple settings.
- Prompt and support participants to interpret their learning experiences in light of relevant prior knowledge, experiences, and interests.
- Support and encourage learners to extend their learning over time.

Recommendation 2: From their inception, informal environments for science learning should be developed through community–educator partnerships and whenever possible should be rooted in scientific problems and ideas that are consequential for community members.

Recommendation 3: Educational tools and materials should be developed through iterative processes involving learners, educators, designers, and experts in science, including the sciences of human learning and development.

Front-Line Educators

Front-line educators include the professional and volunteer staff of institutions and programs that offer and support science learning experiences. In some ways, even parents and other care providers who interact with learners in these settings are front-line educators. Front-line educators may model desirable science learning behaviors, helping learners develop and expand scientific explanations and practice and in turn shaping how learners interact with science, with one another, and with educational materials. They may also serve as the interface between informal institutions and programs and schools, communities, and groups of professional educators. Given the diversity of community members who do (or could) participate in informal environments, front-line educators should embrace diversity and work thoughtfully with diverse groups.

Recommendation 4: Front-line staff should actively integrate questions, everyday language, ideas, concerns, world views, and histories, both their own and those of diverse learners. To do so they will need support opportunities to develop cultural competence, and to learn with and about the groups they want to serve.

Researchers and Evaluators

Improving the quality of evidence on learning science in informal environments is a paramount challenge. Research and evaluation efforts rely on partnerships among curators, designers, administrators, evaluators, researchers, educators, and other stakeholders whose varied interests, expertise, and resources support and sustain inquiry. Accordingly our recommendations address investigators and the broader community that collaborates with investigators and consumes research and evaluation results.

Recommendation 5: Researchers, evaluators, and other leaders in informal education should broaden opportunities for publication of peer-reviewed research and evaluation and provide incentives for investigators in non-academic positions to publish their work in these outlets.

Recommendation 6: Researchers and evaluators should integrate bodies of research on learning science in informal environments by developing theory that spans venues and links cognitive, affective, and sociocultural accounts of learning.

Recommendation 7: Researchers and evaluators should use assessment methods that do not violate participants' expectations about learning in informal settings. Methods should address the science strands, provide valid evidence across topics and venues, and be designed in ways that allow educators and learners alike to reflect on the learning taking place in these environments.

———

Taking Science to School: Learning and Teaching Science in Grades K-8 **(2007)**

General Description

What is science for a child? How do children learn about science and how to do science? Drawing on a vast array of evidence from neuroscience to classroom observation, *Taking Science to School* provides a comprehensive picture of what we know about teaching and learning science from kindergarten through eighth grade. This book provides a basic foundation for guiding science teaching and supporting students in their learning developed around three fundamental questions:

    1. How is science learned, and are there critical stages in children's development of scientific concepts?

2. How should science be taught in K-8 classrooms?
3. What research is needed to increase understanding about how students learn science?

This book also offers recommendations on professional development. How science is taught ultimately depends on teachers. Extensive rethinking of how teachers are prepared before they begin teaching and as they continue teaching—and as science changes—is critical to improving K-8 science education.

This book offers a new framework for what it means to be proficient in science. This framework rests on a view of science as both a body of knowledge and an evidence-based, model-building enterprise that continually extends, refines, and revises knowledge. This framework moves beyond a focus on the dichotomy between either content knowledge or process skills because content and process are inextricably linked in science. In this framework, students who are proficient in science:

1. know, use, and interpret scientific explanations of the natural world;
2. generate and evaluate scientific evidence and explanations;
3. understand the nature and development of scientific knowledge; and
4. participate productively in scientific practices and discourse.

Relevance to Convocation Participants

This book is directly relevant to researchers and practitioners alike. (A separate, more practitioner-oriented, volume *Ready, Set, Science!* has been developed based on its content and is described separately in this appendix.) *Taking Science to School* is important because it makes clear that effective science education requires a complex interplay among content knowledge, investigation, reflection, and discourse skills. Development of these skills is influenced by maturation, experience, prior instruction, and opportunities to learn as well as gender, ethnicity, socioeconomic status, and cultural experiences. Research makes clear that many children arrive at school already able to understand concepts and think at a level that many educators had previously thought to be impossible. Younger students are capable of engaging with concepts and interacting with each other in ways that allow them to develop all four strands of scientific capability. The book calls on the education community to rethink what constitutes effective science education for all children based on the growing body of knowledge in human learning generally and early childhood learning and education more specifically.

Recommendations

Recommendation 1: Developers of standards, curriculum, and assessment should revise their frameworks to reflect new models of children's thinking and take better advantage of children's capabilities.

Recommendation 2: The next generation of standards and curricula at both the national and state levels should be structured to identify a few core ideas in a discipline and elaborate how those ideas can be cumulatively developed over grades K-8.

Recommendation 3: Developers of curricula and standards should present science as a process of building theories and models using evidence, checking them for internal consistency and coherence, and testing them empirically. Discussions of scientific methodology should be introduced in the context of pursuing specific questions and issues rather than as templates or invariant recipes.

Recommendation 4: Science instruction should provide opportunities for students to engage in all four strands of science proficiency. If these four strands are realized, children will be able to:

1. know, use, and interpret scientific explanations of the natural world;
2. generate and evaluate scientific evidence and explanations;
3. understand the nature and development of scientific knowledge; and
4. participate productively in scientific practices and discourse.

Recommendation 5: State and local leaders in science education should provide teachers with models of classroom instruction that provide opportunities for interaction in the classroom, where students carry out investigations and talk and write about their observations of phenomena, their emerging understanding of scientific ideas, and ways to test them.

Recommendation 6: State and local school systems should ensure that all K-8 teachers experience sustained science-specific professional development in preparation and while in service. Professional development should be rooted in the science that teachers teach and should include opportunities to learn about science, about current research on how children learn science, and about how to teach science.

Recommendation 7: University-based science courses for teacher candidates and teachers' ongoing opportunities to learn science in service should mirror the opportunities they will need to provide for their students, that is, incorporating practices in the four strands that constitute science proficiency and giving sustained attention to the core ideas in the discipline. The topics of study should be aligned with central topics in the K-8 curriculum.

Recommendation 8: Federal agencies that support professional development should require that the programs they fund incorporate models of instruction that combine the four strands of science proficiency, focus on core ideas in science, and enhance teachers' science content knowledge, knowledge of how students learn science, and knowledge of how to teach science.

Ready, Set, Science! Putting Research to Work in K-8 Science Classrooms **(2007)**

General Description

What types of instructional experiences help K-8 students learn science with understanding? What do science educators teachers, teacher leaders, science specialists, professional development staff, curriculum designers, school administrators need to know to create and support such experiences?

Directed toward education practitioners and filled with classroom case studies that bring to life research findings and help readers to replicate success, *Ready, Set, Science!* provides an overview of the groundbreaking and comprehensive synthesis of research into teaching and learning science in kindergarten through eighth grade that is detailed in *Taking Science to School: Learning and Teaching Science in Grades K-8.* This practitioner-oriented volume summarizes a rich body of findings from the learning sciences and presents detailed cases of science educators at work to make the implications of research clear, accessible, and stimulating for a broad range of science educators. It richly illustrates the four strands of learning that are featured in *Taking Science to School,* i.e., that children will be able to:

1. know, use, and interpret scientific explanations of the natural world;
2. generate and evaluate scientific evidence and explanations;
3. understand the nature and development of scientific knowledge; and
4. participate productively in scientific practices and discourse.

The examples presented are based on real classroom experiences that illustrate the complexities with which teachers grapple every day. They show how expert teachers work to select and design rigorous and engaging instructional tasks, manage classrooms, orchestrate productive discussions with culturally and linguistically diverse groups of students, and help students make their thinking visible using a variety of representational tools.

Relevance to Convocation Participants

There are many reasons that science must be taught and learned in ways that encourage younger children to become interested in this subject area:

1. Science is an enterprise that can be harnessed to improve quality of life on a global scale.
2. Science may provide a foundation for the development of language, logic, and problem-solving skills in the classroom.
3. A democracy demands that its citizens make personal, community-based, and national decisions that involve scientific information.
4. For some students, science will become a lifelong vocation or avocation.

New research points toward a kind of science education that differs substantially from what occurs in most science classrooms today. This new vision of science education embraces different ways of thinking about science, different ways of thinking about students, and different ways of thinking about science education.

Given that a goal of this convocation is to help participants think much more deeply about the role of science education in furthering the development of *all* of California's students in grades K-8, and the interconnections between science and other subject domains, this book offers those who are tasked with implementing this vision with the background and resources they will need to do so effectively.

Report Table of Contents

Because this is a practitioner volume that is a derivative of *Taking Science to School*, it does not include specific conclusions or recommendations. That information can be found in the more technical *Taking Science to School*. Instead this volume contains the following themes organized as chapters and appendices; all sections of the book are directly relevant to this convocation.

Chapters
1. A New Vision of Science in Education
2. Four Strands of Science Learning
3. Foundational Knowledge and Conceptual Change
4. Organizing Science Education Around Core Concepts
5. Making Thinking Visible: Talk and Argument
6. Making Thinking Visible: Modeling and Representation
7. Learning from Science Investigations
8. A System That Supports Science Learning

Appendixes
A Questions for Practitioners
B Assessment Items Based on a Learning Progression for Atomic-Molecular Theory
C Academically Productive Talk
D Biographical Sketches of Oversight Group and Coauthors

Enhancing Professional Development for Teachers: Potential Uses of Information Technology (2007)

General Description

This report is a comprehensive overview of a workshop organized by a committee of teachers and other education experts and hosted by the National Academies Teacher Advisory Council and the California Teacher Advisory Council. The workshop was developed to explore a vision of the potential of online teacher professional development, its challenges, and the research needed to understand and advance this rapidly emerging area. In the workshop presentations and discussions, master classroom teachers joined with researchers, curriculum and information technology developers, professional development experts, state-level policy makers, principals, and foundation representatives. This report is addressed to all of the audiences represented by these participants.

Teachers too often have experienced a "one-size-fits-all" professional development model, in which someone else decides what they need to learn. And too often experiences with professional development focus primarily on improvement (i.e., remediation) rather than professional growth and exploration of new ideas, cutting-edge developments in a teacher's field of expertise, or promising new pedagogies. This conventional model seldom meets the particular needs of teachers in specific fields and disciplines, such as mathematics, science, and technology. Recognizing ineffective professional development as a critical issue, the

Teacher Advisory Council convened a workshop in October 2004 and issued a report called *Linking Mandatory Professional Development with High-Quality Teaching and Learning* (National Research Council, 2006, see http://www.nap.edu/catalog.php?record_id=11518).

In an effort to build on the knowledge gained at the 2004 workshop, the Teacher Advisory Council began to explore emerging opportunities in professional development. Council members saw the potential for online learning technologies to provide professional development that could be far more tailored to the needs of science, mathematics, and technology teachers, to all teachers at different stages of their professional careers, and to teachers located in places where access to high-quality face-to-face professional development experiences to their schools is difficult.

Relevance to Convocation Participants

An accumulating body of evidence is showing that effective teachers are one of the most important contributors to science learning. Professional development designed for teachers of science, especially in the elementary and middle grades, can be integral to improving teachers' classroom practice and to empowering them as professionals. Given the new, expansive array of electronic tools that are being developed to enhance student learning, the participants at this workshop concluded that additional time, effort, and resources should be devoted to studying much more intensively their uses and applications in making quality professional development experiences available to teachers that are more tailored to their individual needs and the stages of their careers in teaching. The hardware and software industries in California, working in concert with education researchers, professional development providers, and classroom teachers, could make significant advances in this realm of education.

Report Table of Contents

Because this is a report from a workshop, it contains no recommendations.

Obstacles to Online Teacher Professional Development
  Lack of Knowledge
  Lack of Support from Administrators
  Lack of Access to Technologies
  Lack of Time, Financial, and Parental Support
  Lack of Materials
  Lack of Support from Higher Education
  Changing Teachers' Beliefs and Practices
Teacher Leadership
The Need for Research on Online Teacher Professional Development
Next Steps
  Providing Teachers, Administrators, and Policy Makers with
    Information
  Building Support Among Administrators and Policy Makers
  Providing Teachers with Access to Online Technologies
  Fostering Development of Good Materials
  Changing Teachers' Beliefs and Practices
  Involving Teachers as Active Participants in Planning and
    Implementation
Appendixes
  A Workshop Agenda and Participants
  B Workshop Materials
  C Programs Highlighted During the Workshop
  D Biographical Sketches of Committee Members and Workshop
    Presenters

———

Tech Tally: Approaches to Assessing Technological Literacy (2006)

General Description

In a broad sense, technology is any modification of the natural world made to fulfill human needs or desires. Although people tend to focus on the most recent technological inventions, technology includes a myriad of devices and systems that profoundly affect everyone in modern society. Technology is pervasive; an informed citizenry needs to know what technology is, how it works, how it is created, how it shapes our society, and how society influences technological development. This understanding depends in large part on an individual level of technological literacy.

No one really knows the level of technological literacy among people in this country—or for that matter, in other countries. Although many concerns have been raised that Americans are not as technologically literate as they should be, these statements are based on general impressions with little hard data to back them up. Therefore, the starting point for improving technological literacy must be to determine the current level of technological understanding and capability, areas that require improvement first, and how technological literacy varies among different populations—children and adults, for instance.

Tech Tally: Approaches to Assessing Technological Literacy uses the metaphor of design to talk about how an assessment of technological literacy might be constructed, includes a primer on educational assessment issues for nonexperts, and reviews several dozen assessment instruments that directly or indirectly measure some aspect of technological literacy for students, teachers, or out-of-school adults. The book also:

- examines opportunities and obstacles to developing scientifically valid and broadly applicable assessment instruments for technological literacy in the three target populations;
- includes several sample case studies, one at the state level, of how assessment in this domain might be done;
- proposes an assessment matrix that suggests how content areas and cognitive domains of technology might be accounted for in an assessment of technological literacy; and
- explores computer-based assessment methods that might be particularly suited to assessment of technological literacy.

Relevance to Convocation Participants

If technology and technological literacy are to become meaningful components of science education in California and elsewhere, attention must be paid to assessment issues. As *Tech Tally* makes clear, however, assessment in this domain poses significant challenges, and to date there are no ready-made assessment instruments available. One chapter of the report and a related recommendation suggest the potential value of computer-based assessment methods. Given the concentration of academic and industrial activity in California related to computer technology and software, research on computer-based assessment for achievement in science, technology, engineering, and mathematics (STEM) education could be a valuable component of the overall planning. Finally, in response to *Tech Tally*, the National Assessment Governing Board (NAGB, the overseers of the National Assessment of Education Progress) has begun a feasibility study of the assessment of technological literacy. The pilot will be fielded in 2012 and, depending on the result, NAGB may add an assessment of technological literacy to its portfolio of national and state tests. This addition could have potentially important implications for science education in California.

Recommendations

Recommendation 1: The National Assessment Governing Board, which oversees the National Assessment of Educational Progress (NAEP), should authorize special studies of the assessment of technological literacy as part of the 2009 NAEP mathematics and science assessments and the 2010 NAEP U.S. history assessment. The studies should explore the content connections between technology, science, mathematics, and U.S. history to determine the feasibility of adding technology-related items to future NAEP assessments in these subjects.

Recommendation 2: The U.S. Department of Education and National Science Foundation should send a recommendation to the International Association for the Evaluation of Educational Achievement and the Trends in Mathematics and Science Study (TIMSS) governing board encouraging them to include technological literacy items in TIMSS assessments as a context for assessments of science and mathematics. The U.S. Department of Education and National Science Foundation should send a recommendation to the Organization for Economic Cooperation and Development and the governing board for the Programme for International Student Assessment (PISA) supporting the inclusion of technological literacy items as a cross-curricular competency.

Recommendation 3: The National Science Foundation should fund a number of sample-based studies of technological literacy in K-12 students. The studies should have different assessment designs and should assess different population subsets, based on geography, population density, socioeconomic status, and other factors. Decisions about the content of test items, the distribution of items among the three dimensions of technological literacy, and performance levels should be based on a detailed assessment framework.

Recommendation 4: When states determine whether teachers are "highly qualified" under the provisions of the No Child Left Behind Act (NCLB), they should ensure—to the extent possible—that assessments used for this purpose include items that measure technological literacy. This is especially important for science, mathematics, history, and social studies teachers, but it should also be considered for teachers of other subjects. In the review of state plans for compliance with NCLB, the U.S. Department of Education should consider the extent to which states have fulfilled this objective.

Recommendation 5: The National Science Foundation and U.S. Department of Education should fund the development and pilot testing of sample-based assessments of technological literacy among preservice and inservice teachers of science, technology, English, social studies, and mathematics. These assessments should be informed by carefully developed assessment frameworks. The results should be disseminated to schools of education, curriculum developers, state boards of education, and other groups involved in teacher preparation and teacher quality.

Recommendation 6: The International Technology Education Association should continue to conduct a poll on technological literacy every several years, adding items that address the three dimensions of technological literacy, in order to build a database that reflects changes over time in adult knowledge of and attitudes toward technology. In addition, the U.S. Department of Education, working with its international partners, should expand the problem-solving component of the Adult Literacy and Life Skills Survey to include items relevant to the assessment of technological literacy. These items should be designed to gauge participants' general problem-solving capabilities in the context of familiar, relevant situations. Agencies that could benefit by knowing more about adult understanding of technology, such as the National Science Foundation, U.S. Department of Education, U.S. Department of Defense, and National Institutes of Health, should consider funding projects to develop and conduct studies of technological literacy. Finally, opportunities for integrating relevant knowledge and attitude measures into existing studies, such as the General Social Survey, the National Household Education Survey, and Surveys of Consumers, should be pursued.

Recommendation 7: The National Science Foundation or U.S. Department of Education should fund a synthesis study focused on how children learn technological concepts. The study should draw on the findings of multidisciplinary research in mathematics learning, spatial reasoning, design thinking, and problem solving. The study should provide guidance on pedagogical, assessment, teacher education, and curricular issues of interest to educators at all levels, teacher-education providers and licensing bodies, education researchers, and federal and state education agencies.

Recommendation 8: The National Science Foundation (NSF) and U.S. Department of Education should support a research-capacity-building initiative related to the assessment of technological literacy. The initiative should focus on supporting graduate and postgraduate research related to how students and teachers learn technology and engineering concepts. Funding should be directed to academic centers of excellence in education research—including, but not limited to, NSF-funded centers for learning and teaching—whose missions and capabilities are aligned with the goal of this recommendation. To the committee's knowledge, no rigorous efforts have been made to ascertain how adults acquire and use technological knowledge. School and work experience could affect their performance, but adults who are no longer in the formal education system are also influenced by a variety of free-choice learning opportunities, including popular culture, the news media, and museums and science centers.

Recommendation 9: The National Science Foundation should take the lead in organizing an interagency federal research initiative to investigate technological learning in adults. Because adult learning is continuous, longitudinal studies should be encouraged. Informal learning institutions that engage broad populations, such as museums and science centers, should be considered important venues for research on adult learning, particularly related to technological capability. To ensure that the perspectives of adults from a variety of cultural and socioeconomic backgrounds are included, studies should also involve community colleges, nonprofit community outreach programs, and other programs that engage diverse populations.

Recommendation 10: The National Institute of Standards and Technology, which has a broad mandate to promote technology development and an extensive track record in organizing research conferences, should convene a major national meeting to explore the potential of innovative, computer-based techniques for assessing technological literacy in students, teachers, and out-of-school adults. The conference should be informed by research related to assessments of science inquiry and scientific reasoning and should consider how innovative assessment techniques compare with traditional methods.

Recommendation 11: Assessments of technological literacy in K-12 students, K-12 teachers, and out-of-school adults should be guided by rigorously developed assessment frameworks, as described in this report.

- **For K-12 students,** the National Assessment Governing Board, which has considerable experience in the development of assessment frameworks in other subjects, should commission the development of a framework to guide the development of national and state-level assessments of technological literacy.
- **For K-12 teachers,** the National Science Foundation and U.S. Department of Education, which both have programmatic interests in improving teacher quality, should fund research to develop a framework for an assessment of technological literacy in this

population. The research should focus on (1) determining how the technological literacy needs of teachers differ from those of student populations and (2) strategies for implementing teacher assessments in a way that would provide useful information for both teachers and policy makers. The resulting framework would be a prerequisite for assessments of all teachers, including generalists and middle- and high-school subject-matter specialists.

- **For out-of-school adults,** the National Science Foundation and U.S. Department of Education, which both have programmatic activities that address adult literacy, should fund research to develop a framework for the assessment of technological literacy in this population. The research should focus on determining thresholds of technological literacy necessary for adults to make informed, everyday, technology-related decisions.

Recommendation 12: The U.S. Department of Education, state education departments, private educational testing companies, and education-related accreditation organizations should broaden the definition of "technological literacy" to include not only the use of educational technologies (computers) but also the study of technology, as described in the International Technology Education Association *Standards for Technological Literacy* and the National Academy of Engineering and National Research Council report *Technically Speaking*.

———

Technically Speaking: Why All Americans Need to Know More About Technology (2002)

General Description

The United States is riding a whirlwind of technological change. To be sure, there have been periods, such as the late 1800s, when new inventions appeared in society at a comparable rate. But the pace of change today, and its social, economic, and other impacts, are as significant and far-reaching as at any other time in history. And it seems that the faster we embrace new technologies, the less we are able to understand them. What is the long-term effect of this galloping technological revolution? In today's world, it is nothing less than a matter of responsible citizenship to grasp the nature and implications of technology.

Technically Speaking provides a blueprint for bringing us all up to speed on the role of technology in our society, including understanding such distinctions as technology versus science and technological literacy versus technical competence. It explains what it means to be a technologically literate citizen. The book goes on to explore the social, historical, political, and educational contexts of technological literacy.

This overview highlights specific issues of concern: the state of technological studies in K-12 schools, the reach of the Internet into people's homes and lives, and the crucial role of technology in today's economy and workforce. Three case studies, related to car airbags, genetically modified foods, and the 2001 California energy crisis, illustrate why ordinary citizens need to understand technology to make responsible decisions.

Relevance to Convocation Participants

Technically Speaking is relevant to this convocation because it clearly explains and makes the case for the "T" in STEM (science, technology, engineering, and mathematics) education. As the report points out, policy makers, educators, and the public alike tend to think of technology quite narrowly, as either the use of computers and other electronics or as educational technology—a tool for classroom learning. This vision of technology is quite limited in scope, as *Technically Speaking* makes clear. A broader view of technology, as the human-constructed world, is consistent with how engineers and scientists see the world and gives technology equal-partner status within the STEM quartet of subjects. The report also defines and presents a conceptual model for "technological literacy," a quality that captures a complex mix of knowledge, capability, and ways of thinking and acting.

Recommendations

Recommendation 1: Federal and state agencies that help set education policy should encourage the integration of technology content into K-12 standards, curricula, instructional materials, and student assessments in non-technology subject areas.

Recommendation 2: The states should better align their K-12 standards, curriculum frameworks, and student assessment in the sciences, mathematics, history, social studies, civics, the arts, and language arts with national educational standards that stress the connections between these subjects and technology. National Science Foundation (NSF)- and Department of Education (DoEd)-funded instructional materials and informal-education initiatives should also stress these connections.

Recommendation 3: NSF, DoEd, state boards of education, and others involved in K-12 science education should introduce, where appropriate, the word "technology" into the titles and contents of science standards, curricula, and instructional materials.

Recommendation 4: NSF, DoEd, and teacher education accrediting bodies should provide incentives for institutions of higher education to transform the preparation of all teachers to better equip them to teach about technology throughout the curriculum.

Recommendation 5: The National Science Foundation should support the development of one or more assessment tools for monitoring the state of technological literacy among students and the public in the United States.

Recommendation 6: The National Science Foundation and the Department of Education should fund research on how people learn about technology, and the results should be applied in formal and informal education settings.

Recommendation 7: Industry, federal agencies responsible for carrying out infrastructure projects, and science and technology museums should provide more opportunities for the nontechnical public to become involved in discussions about technological developments.

Recommendation 8: Federal and state government agencies with a role in guiding or supporting the nation's scientific and technological enterprise, and private foundations concerned about good governance, should support executive education programs intended to increase the technological literacy of government and industry leaders.

Recommendation 9: U.S. engineering societies should underwrite the costs of establishing government- and media-fellow programs with the goal of creating a cadre of policy experts and journalists with a background in engineering.

Recommendation 10: The National Science Foundation, in collaboration with industry partners, should provide funding for awards for innovative, effective approaches to improving the technological literacy of students or the public at large.

Recommendation 11: The White House should add a Presidential Award for Excellence in Technology Teaching to those that it currently offers for mathematics and science teaching.

———

Adding It Up: Helping Children Learn Mathematics (2001)

General Description

The first two paragraphs of the executive summary for *Adding it Up* provide a clear and compelling rationale of new ways of thinking about and learning mathematics for younger children (p. 1):

> Mathematics is one of humanity's great achievements. By enhancing the capabilities of the human mind, mathematics has facilitated the development of science, technology, engineering, business, and government. Mathematics is also an intellectual achievement of great sophistication and beauty that epitomizes the power of deductive reasoning. For people to participate fully in society, they must know basic mathematics. Citizens who cannot reason mathematically are cut off from whole realms of human endeavor. Innumeracy deprives them not only of opportunity but also of competence in everyday tasks.
>
> The mathematics students need to learn today is not the same mathematics that their parents and grandparents needed to learn. When today's students become adults, they will face new demands for mathematical proficiency that school mathematics should attempt to anticipate. Moreover, mathematics is a realm no longer restricted to a select few. *All young Americans must learn to think mathematically, and they must think mathematically to learn.*

Adding It Up explores how students in pre-K through eighth grade learn mathematics and recommends how teaching, curricula, and teacher education should change to improve mathematics learning during these critical years.

The committee identified five interdependent components of mathematical proficiency and described how students develop this proficiency, all of which must be woven together and interconnected (see the braid metaphor on p. 126). They include

Conceptual understanding—comprehension of mathematical concepts, operations, and relations;

Procedural fluency—skill in carrying out procedures flexibly, accurately, efficiently, and appropriately;

Strategic competence—ability to formulate, represent, and solve mathematical problems;

Adaptive reasoning—capacity for logical thought, reflection, explanation, and justification; and

Productive disposition—habitual inclination to see mathematics as sensible, useful, and worthwhile, coupled with a belief in diligence and one's own efficacy.

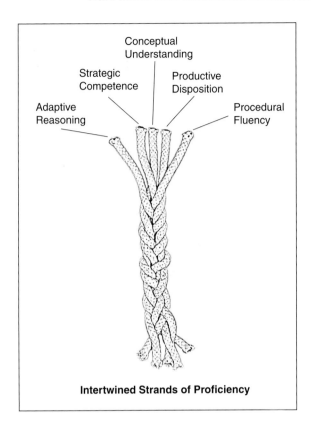

Intertwined Strands of Proficiency

Relevance to Convocation Participants

Just as *Taking Science to School* emphasizes the connections among different "strands" in science education, *Adding It Up* emphasizes the coherence and interconnectedness of mathematical concepts for children and how they might be taught most effectively. Because the concept of number is used so broadly for mathematics education in the early grades, this book also focuses on that concept but shows how students can be provided with rich experiences in mathematics that build on this basis concept that goes well beyond the operational aspects of arithmetic that often are the primary components of mathematics education in the elementary and early middle grades. For example, students can understand that it is possible to communicate about numbers through graphical representations and systems of notation. Students can appreciate and understand that numbers and the operations that are typically emphasized in elementary school mathematics are organized as number systems, such as the whole numbers, and there are regularities of each system that can be discerned.

Numerical computations require algorithms—step-by-step procedures for performing the computations, that can be more or less useful to students depending on how it works and how well it is understood. And finally, the domain of number both supports and is supported by other branches of mathematics, including algebra, measure, space, data, and chance.

Adding It Up also was the first of a series of seminal reports from the National Academies that used knowledge and evidence from the growing research base on human learning and cognition (as described in the NRC report *How People Learn: Brain, Mind, Experience, and School*—see separate overview) as the basis for its findings, conclusions, and recommendations. Others include *Taking Science to School*; *Ready, Set, Science!* (see separate summaries of these books); and *Learning and Understanding: Improving Advanced Study of Mathematics and Science in U.S. High Schools* (2002). The general agreement of the recommendations in this and those subsequent reports strongly suggests that there are fundamental approaches to improving education that are supported by a growing body of research. While learning in specific disciplines may require somewhat different approaches (and there is now a growing area called discipline-based education research), there are also fundamental principles for improving aspects of education, such as teacher professional development, that appear to transcend disciplines.

Recommendations

The overriding premise of *Adding It Up* is that throughout the grades from pre-K through 8 *all* students can and should be mathematically proficient.

Recommendation 1: The integrated and balanced development of all five strands of mathematical proficiency (conceptual understanding, procedural fluency, strategic competence, adaptive reasoning, and productive disposition) should guide the teaching and learning of school mathematics. Instruction should not be based on extreme positions that students learn, on one hand, solely by internalizing what a teacher or book says or, on the other hand, solely by inventing mathematics on their own.

Recommendation 2: Teachers' professional development should be high quality, sustained, and systematically designed and deployed to help all students develop mathematical proficiency. Schools should support, as a central part of teachers' work, engagement in sustained efforts to improve their mathematics instruction. This support requires the provision of time and resources.

Recommendation 3: The coordination of curriculum, instructional materials, assessment, instruction, professional development, and school organization around the development of mathematical proficiency should drive school improvement efforts.

Recommendation 4: Efforts to improve students' mathematics learning should be informed by scientific evidence, and their effectiveness should be evaluated systematically. Such efforts should be coordinated, continual, and cumulative.

Recommendation 5: Additional research should be undertaken on the nature, development, and assessment of mathematical proficiency.

Inquiry and the National Science Education Standards: A Guide for Teaching and Learning (2000)

General Description

"Inquiry" refers to the diverse ways in which scientists study the natural world and in which students grasp science knowledge and the methods by which that knowledge is produced. *Inquiry and the National Science Education Standards* offers a practical guide to teaching inquiry and teaching through inquiry, as recommended by the *National Science Education Standards*. This resource can assist educators who must help school boards, parents, and teachers understand the nature and process of inquiry in science education and the kinds of resources that are required to sustain and nurture it.

This book explains and illustrates how inquiry helps students learn science content, master how to do science, and understand the nature of science. It explores the dimensions of teaching and learning science as inquiry for K-12 students across a range of science topics. Detailed examples help clarify when teachers should use the inquiry-based approach and how much structure, guidance, and coaching they should provide for students.

The book dispels myths that may have discouraged some educators from the inquiry-based approach and elucidates the interplay among concepts, processes, and science as it is experienced in the classroom. *Inquiry and the National Science Education Standards* offers a number of classroom vignettes that explore different kinds of inquiries for elementary, middle, and high school. A section of Frequently Asked Questions addresses teachers' common concerns, such as obtaining appropriate supplies for this kind of pedagogy.

In addition, the book discusses why assessment is important, looks at existing schemes and formats, and addresses how to involve students in assessing their own learning. This book also discusses administrative assistance, communication with parents, appropriate teacher evaluation, and other avenues to promoting and supporting this new teaching and learning paradigm.

Relevance to Convocation Participants

Research indicates that multiple approaches to teaching must be employed to truly engage students from diverse backgrounds and levels of interest in this subject area. While the evidence suggests that a variety of inquiry-based approaches to teaching and learning can indeed reach and engage students who otherwise claim to dislike or be uninterested in science, major problems remain with their implementation, including:

- The education community has not settled on common definitions for what constitutes inquiry-based approaches to teaching and learning.
- Many teachers did not experience inquiry-based approaches to science when they themselves were students and thus may not know how to begin to do so. This problem may increase in the higher grades of K-12 education and especially at the postsecondary level, where few faculty have received professional development in its use and implementation.
- To school officials and parents who have not experienced this approach, inquiry-based science may give the impression that little learning is actually occurring. It may be viewed as unstructured and even chaotic to the casual observer. Thus, there are concerns about what students are and are not learning.
- Because of the traditional dichotomy that has existed in the minds of many science educators between content and process in science education, an emphasis on inquiry is viewed by some as a diminution of content at a time when high-stakes tests seem to emphasize mastery of content.
- It is much more difficult and expensive to assess learning through inquiry-based approaches than more traditional routes of teaching.

Thus, this book, written primarily for the practitioner and teacher educator audiences, can provide very helpful insights about ways to implement inquiry-based teaching and learning.

Report Table of Contents

This book is a derivative of and supplement to the National Research Council's *National Science Education Standards*. Thus, there are no specific policy recommendations in this book beyond those in the *National Science*

Education Standards. To give convocation participants a better sense of the scope of *Inquiry and the National Science Education Standards*, the table of contents is reproduced here.

1. Inquiry in Science and in Classrooms
2. Inquiry in the National Science Education Standards
3. Images of Inquiry in K-12 Classrooms
4. Classroom Assessment and Inquiry
5. Preparing Teachers for Inquiry-Based Teaching
6. Making the Case for Inquiry
7. Frequently Asked Questions About Inquiry
8. Supporting Inquiry-Based Teaching and Learning

Appendixes
A Excerpts from the National Science Education Standards
B Selecting Instructional Materials
C Resources for Teaching Science Through Inquiry

How People Learn: Brain, Mind, Experience, and School—Expanded Edition (2000)

General Description

How People Learn offers a detailed review, synthesis, and analysis of exciting research about the brain and human learning that provides answers to a number of compelling questions. When do infants begin to learn? How do experts learn and how is this learning different from that of novices? What happens to how people process information when they move from being a novice to an expert in some subject domain? Does expertise in one subject area allow experts to also better understand other subject areas more quickly? What can teachers and schools do—with curricula, classroom settings, and teaching methods—to help children learn most effectively?

New evidence from many branches of science has significantly added to the understanding of what it means to know, from the neural processes that occur during learning to the influence of culture on what people see and absorb. *How People Learn* examines these findings and their implications for what we teach, how we teach it, and how we assess how our children learn. The book uses exemplary teaching to illustrate how approaches based on what we now know result in in-depth learning. This new knowledge calls into question concepts and practices firmly entrenched in our current education system.

Topics in this book include

- How learning actually changes the physical structure of the brain.
- How existing knowledge affects what people notice and how they learn.
- What the thought processes of experts tell us about how to teach.
- The amazing learning potential of infants.
- The relationship of classroom learning and everyday settings of community and workplace.
- Learning needs and opportunities for teachers.
- A realistic look at the role of technology in education.

Originally released in hardcover in the 1999, *How People Learn* was expanded in 2000 to show how the theories and insights from the original book can translate into actions and practice, thus making a real connection between classroom activities and learning behaviors.

Relevance to Convocation Participants

The rich set of evidence presented, findings, and conclusions in *How People Learn* have direct application and important implications for convocation participants. This book has served as the basis for many subsequent reports in education that have been authored by expert committees of the National Academies.

How People Learn will provide a vitally important guide to the kinds of research evidence in human learning and cognition and can serve as a very useful guide to the leaders of efforts to improve science education in grades K-8 throughout California in ways that are steeped in solid research evidence.

Recommendations

Unlike other NRC studies that include specific recommendations, this committee of experts instead decided to present implications of the research and evidence on human learning for teachers, school, and the larger education system. They offer conclusions that are dispersed throughout the book and too numerous to list in their entirety here. The major set of implications for education is provided below. The book provides much more detail about each of these statements:

Key Findings:

1. Students come to the classroom with preconceptions about how the world works. If their initial understanding is not engaged, they

may fail to grasp the new concepts and information that are taught, or they may learn them for purposes of a test but revert to their preconceptions outside the classroom.

2. To develop competence in an area of inquiry, students must: (a) have a deep foundation of factual knowledge, (b) understand facts and ideas in the context of a conceptual framework, and (c) organize knowledge in ways that facilitate retrieval and application.

3. A "metacognitive" approach to instruction can help students learn to take control of their own learning by defining learning goals and monitoring their progress in achieving them.

Implications for Teaching:

1. Teachers must draw out and work with the preexisting understandings that their students bring with them [pre- and misconceptions].

2. Teachers must teach some subject matter in depth, providing many examples in which the same concept is at work and providing a firm foundation of factual knowledge.

3. The teaching of metacognitive skills should be integrated into the curriculum in a variety of subject areas.

Implications for Designing Classroom Environments:

1. Schools and classrooms must be learner centered.

2. To provide a knowledge-centered classroom environment, attention must be given to what is taught (information, subject matter), why it is taught (understanding), and what competence or mastery looks like.

3. Formative assessments—ongoing assessments designed to make students' thinking visible to both teachers and students—are essential. They permit the teacher to grasp the students' preconceptions, understand where the students are in the "developmental corridor" from informal to formal thinking, and design instruction accordingly. In the assessment-centered classroom environment, formative assessments help both teachers and students monitor progress.

4. Learning is influenced in fundamental ways by the context in which it takes place. A community-centered approach requires the development of norms for the classroom and school, as well as connections to the outside world, that support core learning values.

———

Educating Teachers of Science, Mathematics, and Technology: New Practices for the New Millennium (2000)

General Description

Written by a committee that included experts in science, mathematics, and technology education at both the K-12 and postsecondary levels (including teacher practitioners at various grade levels), *Educating Teachers of Science, Mathematics, and Technology* synthesized what was known at the time about the quality of math and science teaching in the United States, drawing conclusions about why teacher preparation needs to change, and then outlining recommendations for accomplishing those changes.

Educating Teachers addresses the issues associated with teacher education and professional development from a variety of contexts. It begins by helping readers understand the kinds of challenges that teachers often routinely face in their classrooms and in their schools and districts. It compares and contrasts teaching as a profession with other professions in the United States.

The book then synthesizes the research literature about the importance of having effective teachers in classrooms and what might constitute more effective teacher education. One of the important insights from this report is that, rather than being seen as distinctive entities that are controlled and managed very differently from each other, pre- and inservice education of teachers should instead be viewed as a seamless continuum that begins when a student makes decisions about whether to become a teacher (similar to prelaw or premedical advising) through a professional career that progresses from induction to experienced teacher to master teacher. It suggests instead that increased emphasis be placed on career-long *teacher education*.

Also examined are important issues in teacher professionalism: what teachers should be taught about their subjects, the usefulness of inservice education to novice and experienced teachers, the challenge of program funding, and the merits of various kinds of credentialing. Professional development schools are reviewed and vignettes are presented that describe exemplary teacher development practices.

As a framework for addressing the task of revamping teacher education, the book (and especially Chapter 6) offers a vision for fundamentally different relationships than currently exist among most school districts, two- and four-year colleges, and universities. It also offers recommendations about how teachers can experience professional growth throughout their careers that may help them stay in the classroom rather than feeling compelled to move to other areas of education (e.g., school administration) in order to advance in their careers.

Relevance to the California STEM Innovation Network

The ultimate success of the California STEM Education Network is likely to be determined to a large extent by the time, resources, and knowledge base that are devoted to preparing California's current workforce of teachers of science and mathematics to approach their craft differently. It also will be determined in part by the ability of the network to (1) convince postsecondary faculty (both in the STEM disciplines and in schools of education) and the leaders of colleges and universities that teacher education is an inherent and primary responsibility of those institutions and (2) to work in partnership with higher education to foster more effective education of teachers at all levels of the professional continuum.

Layered upon the well-known issues surrounding teacher education in science, mathematics, and technology are the issues of

1. preparing both future and currently practicing teachers to focus on new approaches and techniques to promote more inquiry by students in classrooms and laboratories (see also the overviews of *Ready, Set, Science!*; *Adding It Up*; and *Inquiry and the National Science Education Standards*), especially when too few teachers have personally experienced these approaches to STEM education when they were students, and
2. developing new approaches to the continuum of teacher education that helps future and currently practicing teachers more deeply understand the interconnections among the STEM disciplines and how those interconnections can be used to approach teaching and learning of these disciplines in fundamentally different ways. This challenge is likely to be especially difficult because most teachers have not had courses or professional development experiences that help to interconnect science and mathematics, let alone the additional connections of technology and engineering.

Educating Teachers of Science, Mathematics, and Technology can provide the leadership of the California STEM Education Network with new perspectives about approaches to teacher education and how new relationships among the various stakeholders might be developed and nurtured.

Recommendations

General Recommendations

1. Teacher education in science, mathematics, and technology [should] be viewed as a continuum of programs and professional experiences that enables individuals to move seamlessly from college preparation for teaching to careers in teaching these subject areas.

2. Teacher education [should] be viewed as a career-long process that allows teachers of science, mathematics, and technology to acquire and regularly update the content knowledge and pedagogical tools needed to teach in ways that enhance student learning and achievement in these subjects.
3. Teacher education [should] be structured in ways that allow teachers to grow individually in their profession and to contribute to the further enhancement of both teaching and their disciplines.

Specific Recommendations

For Governments:

Local, state, and federal governments should recognize and acknowledge the need to improve teacher education in science and mathematics, as well as assist the public in understanding and supporting improvement. Governments should understand that restructuring teacher education will require large infusions of financial support and make a strong commitment to provide the direct and indirect funding required to support local and regional partnerships for improving teacher education in these disciplines. They also should encourage the recruitment and retention of teachers of science and mathematics—particularly those who are "in-field"—through financial incentives, such as salaries that are commensurate and competitive with those in other professions in science, mathematics, and technology; low-interest student loans; loan forgiveness for recently certified teachers in these disciplines who commit to teaching; stipends for teaching internships; and grants to teachers, school districts, or teacher education partnerships to offset the costs of continual professional development.

For Collaboration Between Institutions of Higher Education and the K-12 Community:

Two- and four-year institutions of higher education and school districts that are involved with partnerships for teacher education should—working together—establish a comprehensive, integrated system of recruiting and advising people who are interested in teaching science, mathematics, and technology.

For the Higher Education Community:

1. Science, mathematics, and engineering departments at two- and four year colleges and universities should assume greater responsibility for offering college-level courses that provide teachers with strong exposure to appropriate content and that model the kinds of pedagogical approaches appropriate for teaching that content.

2. Two- and four-year colleges and universities should reexamine and redesign introductory college-level courses in science and mathematics to better accommodate the needs of practicing and future teachers.

3. Universities whose primary mission includes education research should set as a priority the development and execution of peer-reviewed research studies that focus on ways to improve teacher education, the art of teaching, and learning for people of all ages. New research that focuses broadly on synthesizing data across studies and linking it to school practice in a wide variety of school settings would be especially helpful to the improvement of teacher education and professional development for both prospective and experienced teachers. The results of this research should be collated and disseminated through a national electronic database or library.

4. Two- and four-year colleges and universities should maintain contact with and provide guidance to teachers who complete their preparation and development programs.

5. Following a period of collaborative planning and preparation, two- and four-year colleges and universities in a partnership for teacher education should assume primary responsibility for providing professional development opportunities to experienced teachers of science, mathematics, and technology. Such programs would involve faculty from science, mathematics, and engineering disciplines and from schools of education.

For the K-12 Education Community:

1. Following a period of collaborative planning and preparation, school districts in a partnership for teacher education should assume primary responsibility for providing high-quality practicum experiences and internships for prospective teachers.

2. School districts in a partnership for teacher education should assume primary responsibility for developing and overseeing field experiences, student teaching, and internship programs for new teachers of science, mathematics, and technology.

3. School districts should collaborate with two- and four-year colleges and universities to provide professional development opportunities to experienced teachers of science, mathematics, and technology. Such programs would involve faculty from science, mathematics, and engineering disciplines and from schools of education. Teachers who participate in these programs would, in turn, offer their expertise and guidance to others involved with the partnership.

For Professional and Disciplinary Organizations:

1. Organizations that represent institutions of higher education should assist their members in establishing programs to help new teachers. For example, databases of information about new teachers could be developed and shared among member institutions so that colleges and universities could be notified when a newly certified teacher was moving to their area to teach. Those colleges and universities could then plan and offer welcoming and support activities, such as opportunities for continued professional and intellectual growth.
2. Professional disciplinary societies in science, mathematics, and engineering, higher education organizations, government at all levels, and business and industry should become more engaged as partners (as opposed to advisors or overseers) in efforts to improve teacher education.
3. Professional disciplinary societies in science, mathematics, and engineering, and higher education organizations also should work together to align their policies and recommendations for improving teacher education in science, mathematics, and technology.